MW00816954

MOBILE
GASOLINE

...IT'S
USA

Mobiloil
CHAMPION
SPEEDWAY
CALIFORNIA
UNION OIL
CREDIT CARDS
Honored Here for
MOTOR MAINTENANCE
AND SERVICE
Sky Chief
GASOLINE
GENUINE
Ford
PARTS
AUT
CLEAN
REST
ROOMS
Genuine Ford Parts
For Sale Here
Ford
MAIL POUCH

HOT ROD GARAGES

PETER VINCENT

DEDICATION

This book is dedicated to four friends—Walter Cotten, Pat Foster, Jerry Helwig,

and Dale Poore—all of whom made a difference in my life.

First published in 2009 by Motorbooks, an imprint of MBI Publishing Company, 400 First Avenue North, Suite 400, Minneapolis, MN 55401 USA

ISBN-13: 978-0-7603-4420-0

The Library of Congress has cataloged the first edition of this book as follows:

Vincent, Peter, 1945-
Hot rod garages / Peter Vincent.
p. cm.
Includes index.
ISBN 978-0-7603-2696-1 (hb w/ jkt)
1. Automobile repair shops. 2. Garages. 3. Hot rods—Maintenance and repair. I. Title.
TL153.V57 2009
629.28'7860973--dc22
2008032759

Front cover: Anonymous garage door photographed somewhere in western Idaho.

Frontispiece: Detail, Pete Eastwood's garage, Pasadena, California.

On the title pages: Pokey and Rusty with Phil Linhares' '34 Fordor, Vern Tardel's, Santa Rosa, California.

About the Author
Peter Vincent's work has been featured in more than 20 galleries around the United States and has been published in the *Rodder's Journal*, *American Rodder*, *Hop Up*, *Street Rodder*, and *Rod & Custom*. Vincent, the author and photographer of Motorbooks' *Hot Rod: An American Original* and *Hot Rod: The Photography of Peter Vincent*, lives in Moscow, Idaho.

Acquisition and Project Editor: Dennis Pernu
Content Editor: Peter Bodensteiner
Cover and Interior Designer: John Barnett/4 Eyes Design

Printed in China

10 9 8 7 6 5 4 3 2 1

CONTENTS

FOREWORD

THE SCENE is happy hour at the hotel bar on the Pomona Fairplex grounds. January 2006. Builders, car owners, and vendors in town for the Grand National Roadster Show are taking a break; overhead, flat screens are broadcasting a Barrett-Jackson auction, and a $4.1 million bid has just been posted for a massive General Motors Futureliner. In between cheers and exclamations, the conversation is serious.

The subject is the future—more specifically, what's ahead for hot rod and custom car builders and aftermarket parts manufacturers. The first wave of Baby Boomers is retiring, having prospered and nurtured their offspring to independence; propelled by nostalgia, they now want the cool cars they've dreamed about for decades.

Fast-forward to 2008, and it's happening—the quest for vintage bodies and authentic parts is escalating. Things are getting pricey. The industry has responded with a spectacular array of reproduction parts, and legions of skilled craftsmen are rising to the challenge. Respected, experienced builders across the country are busy producing examples of highly imaginative, superbly crafted cars. Blackie Gejeian's sage advice, "Do it right the first time," is what these builders do for the rest of us who want our next—maybe our last—hot rod done better than the one we built years ago in the backyard.

Some of us who grew up in the backyard-hot-rod-building era have seen gradual evolution toward professionalism in the field, resulting in advances in design and technology. For those of us who remember the old Oakland Roadster Show, with cars placed on hay bales and oil drums serving as stanchions, today's big fairground meets of more than 3,000 cars, the acceptance of hot rods and customs at the Pebble Beach Concours d'Elegance, and the spread of hot rodding to Japan and Europe tell us how far our sport/hobby/pastime has come. Serious students of automotive history now delve deeply into the rich subculture of hot rodding. Significant early cars are researched, properly restored, and placed in public view. The Doane Spencer roadster, the *Beatnik Bandit*, and *The Ala Kart* have become icons, essential images fixed in the mind of every enthusiast.

Peter Vincent has recorded in visual terms a diverse array of builders, both professional shops staffed with skilled specialists, and the privateers, equally skilled builders who may work alone or with a few associates, with production limited in numbers. Vincent has been a fixture on the salt at Bonneville for nearly two decades; his photographs not only document the racecars and their builders, but place them within an environment of astounding beauty.

In this, the third major publication of his work, Peter Vincent has been welcomed into the garages of the many builders he has known for years in order to apply his special vision in a new setting.

Peter has invited the contributions of Michael Dobrin, a veteran automotive historian, and Kevin Thomson, a representative of a younger generation of knowledgeable writers. Dobrin, who worked with Wally Parks in the early days of the National Hot Rod Association, has observed the goings-on in the Oakland, California, shop of Steve Moal and

Phil Linhares with his Vern Tardel–built '34.

has published accounts of many of Moal's unique coachbuilt vehicles. Thomson is seen annually at Bonneville, plying the pits in various Mopar project cars, and has been on the scene through the Tardel/McKenzie No. 917 car's campaign. Here, he covers the prolific work of Vern and Keith Tardel.

This selection of photographs shows us a new dimension in Peter Vincent's art; while his earlier published work concentrated mainly on individual cars on the Bonneville Salt Flats, Muroc, and other sites, in this portfolio Vincent undertakes the challenge of portraiture and interior views. And he portrays the builders and their working environments with a knowing eye and a degree of authenticity that confirms his place as an important observer of this rare segment of American culture.

Philip E. Linhares
Chief Curator of Art
Oakland Museum of California

PREFACE AND ACKNOWLEDGMENTS

A GOOD FRIEND of mine, Clemente Garay, an architect from Spain who studied in the United States, told me that he thought one of the big differences between Europe and the United States was the existence of our garages. These garages are where innovative individuals working on their own have realized some of the country's most significant technological developments.

For the purposes of this book, we are concerned with garages where American hot rods and custom cars are built and restored, although this begs the question of what exactly defines a hot rod or a custom car—or whether there really are definitions. Maybe it's simply a question of our own limitations. Whatever the answer, the shops and garages featured in this book have been the sites of a great variety of automotive creations. Suffice it to say, most of the larger shops featured started in smaller spaces, in some cases even in a home garage. Likewise, some of the home garages featured here have been expanded, while others have remained the same; only the creative processes taking place inside of them have changed or evolved.

I have high respect for everyone included here and only wish I could have featured more. There are many other shops and builders around the country for which I also have an immense amount of admiration.

This has been an interesting book to write, made more complex by a combination of digital photos, film, and black-and-white prints. On top of that, shops changed after they were photographed. I tried to keep up, but at some point I had to call it "good."

First, a special thanks to everybody who gave of their time while I photographed, interviewed, and talked my way through these chapters: Roy Brizio, Pete Eastwood, Cole Foster, Cam Grant, John Gunsaulis, Gary Harms, Terry Hegman, Ron Jolliffe, Bob Lick, Steve Moal, Dick Page, Ken Schmidt and Keith Cornell (The Rolling Bones), Don Small, Cal Tanaka, Vern and Keith Tardel, Bill Vinther, Dale Withers, and Pat Foster, who is missed.

Phil Linhares graciously committed to write the foreword, which I greatly appreciate. I have high respect for his work as chief curator of The Oakland Museum of California. He knows hot rods and owns two that he drives extensively.

Michael Dobrin wrote an important piece for this book, detailing the history of Moal Coachworks. He is quite familiar with the Moal operation and knows of what he writes. Thanks Michael.

Kevin Thomson, with whom I always enjoy working, contributed his reflections of Vern Tardel and the addition of Keith Tardel's Rex Rod operation to Vern's shop, which is a place where I always enjoy taking photographs and spending time. Brett Reed, who now works for Moal Coachworks, also joined me for the trip out to Vern's.

Thanks to Ken Schmidt for all his help and effort on the Rolling Bones chapter. The build-up photographs are his, as is most of the text in that chapter.

Dennis Pernu, Motorbooks acquisitions and project editor—thanks for your patience and help in putting all of this together. Also, thanks to my old friend Peter Bodensteiner, who helped with the text editing.

Finally, congratulations to Kim on your first ride at Bonneville, and in a '32 Ford roadster no less (thanks, Ken). I'm glad you like all this stuff. And to Nathan—thanks for the welcome company, the support, and all your help through the many photography sessions.

I have two projects in my own garage. The '40 Ford coupe is getting a 327, while the '71 Camaro RS is getting a 350 and a Doug Nash five-speed. I am seriously beginning to think of economy with the '40 (i.e., a T5 overdrive tranny and a tall-geared rearend). There is no way of backing off the Camaro engine, though. A just-sold '78 Shovelhead bobber will help finance both cars.

589 BWM

Kurt Lemmon, a talented custom painter and car restorer here in Moscow, Idaho, has helped me extensively with the Camaro bodywork. The car used to go to Bonneville regularly, and the salt took its toll. It will never be a show car, but it's now solid and totally functional. Here it is in Kurt's paint booth. I've owned the car since 1982, and I believe Trans Am racing–style cars are ripe and ready. I am just as excited about this car as I am the '40—and I love '40 coupes.

My unfinished backyard and the corrugated-steel 24×28 insulated garage, which has 10-foot ceilings, lots of light, and attic storage.

Preface and Acknowledgments

My '40 in front of the house. Bob Lick has been a great help with the '40, as have Don Small and Jim Lindsey.

Ford

1
ROY BRIZIO

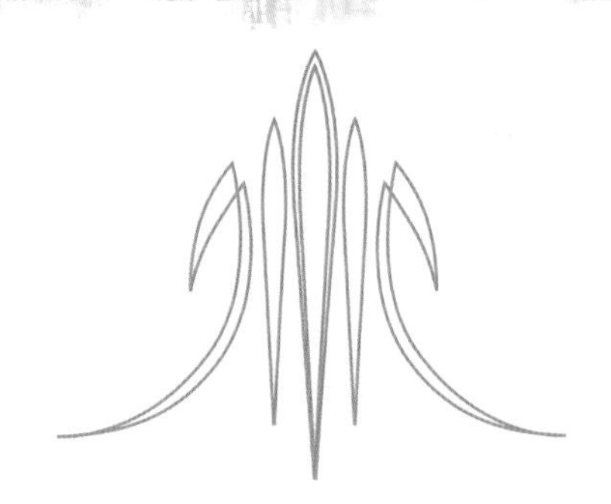

I STARTED NOTICING Roy Brizio's cars in the mid- to late 1980s. Brizio cars have always been noted for their build and design quality, but the car that really brought Roy Brizio Street Rods to my attention was a yellow '33 three-window Ford built for Cliff Hansen. The car was supposed to debut at Pleasanton in 1993, but there was trouble getting it to run correctly at the fairgrounds. What impressed me was that Roy was there with the car, working out the problem. He didn't leave until it was right. Subsequently, I kept noticing the quality of his work remained at a high level.

During the 1990s, magazines seemed to lay claim to certain cars, and editors were reluctant to cover cars that had been featured in other publications. Roy's cars, however, were always very popular with all the publications and usually had been photographed before I ran into them. Many, in fact, were photographed as buildup features.

Roy grew up in a hot rod family. His dad, Andy "The Rodfather" Brizio, ran a hot rod business during the 1960s that specialized in T-bucket chassis. He also ran the Champion Speed Shop in South San Francisco, where Roy worked the counter and developed an ever-increasing passion for hot rods. Andy sold the T-bucket business in 1976, and the following year Roy decided to open his own business at 21 years of age.

To survive the early years, Roy parted out a 1932 three-window, first selling the car's 4-71 blower, then its small-block Chevy, and eventually the fenders and the chassis. The body was the last part to go. Roy sold it to Boyd Coddington.

When Roy Fjastad came out with his '32 Ford frame rails a few years later, Roy started building '32 Ford chassis and the business really started to happen. In 1979, Roy built three turnkey cars: a couple of T-buckets and a Model A roadster. Later that year, former racer Tony Del Rio brought baseball slugger Reggie Jackson over to Roy's shop. Reggie had a commitment with Accel Spark Plugs to build a new car for the SEMA show each year. It was three or four months before SEMA, and Reggie needed a '32 built for the show. Roy had only one other guy working with him at the time and told Reggie that he couldn't do it in that time frame. Reggie persisted, so Roy told him about a fiberglass-bodied '32 roadster they were building that was for sale. Reggie bought it and had Roy repaint the just-painted purple car yellow to match Accel's colors. They finished the car in three months.

The following year, Reggie hired Roy to build him a '32 five-window, and the longtime relationship continued, thanks not only to Roy's talents but also to his integrity and the trust that he builds with his customers. In 1987, Roy won America's Most Beautiful Roadster at the Grand National Roadster Show with James Ells' red Ferrari-powered '32 roadster. From there, his business, always located in South San Francisco, continued to grow, attracting more and more customers looking for quality, turnkey cars. It helped that Roy became friends with his customers during the long build processes.

Roy has built 11 cars for Reggie and 8 cars for Vic Edelbrock Jr., and he has worked with many

A view of the main shop floor from the upper office. Roy is in front of Paulette Zaragoza's '33 Ford Vicky.

The front of Roy Brizio Street Rods. Roy's '40 Ford coupe was a recent find.

figures in the entertainment business, including Jeff Beck, Jimmie Vaughan, and Eric Clapton. In 2006, Roy's shop did some work on a '57 Caddy Biarritz for Neil Young after the car quit near Roy's shop, serendipitously. In 2007, Young, who owns several 1950s automobiles, got in touch with Roy about refurbishing an old '53 Buick Skylark that he has had for some time. By "refurbishing," Young meant mechanical only—no new paint or chrome. Roy explains that Young likes the patina but doesn't want his cars breaking down on him.

Roy moved from his smaller shop on Wattis Way in South San Francisco to the current, much larger digs on Railroad Avenue, also in South San Francisco, in 2002. The new building has truly allowed Roy to spread out, running more consecutive projects with his very talented and dedicated staff.

While at Roy Brizio Street Rods, I saw at least six projects in various stages of completion. One that I watched to completion was the original Sam Barris Mercury custom, in my opinion one of the all-time greats. The Barris restoration was headed by one of Roy's employees, Bill Ganahl, who has a real talent for this type of detail work. He also handled the restoration of George Barris' *Ala Kart* for Roy, another car owned by collector and repeat client John Mumford. I had a chance to see part of John's collection, which is impressive to say the least and very tasteful.

I spent a couple of days photographing and watching work around the shop and talking to Roy about the business. He was a gracious host. I watched him attend to various episodes with different projects in the shop, some of which involved hands-on work. I also witnessed Roy spend time with whomever happened to come into the shop—current, past, and future customers, as well as some who just stopped by to say hello. What I'm getting at is that Roy seems to be in constant motion, dealing with all aspects of the business. You can tell that he loves what he does and truly enjoys the people that he deals with on a day-to-day basis. His impeccable work has earned Roy a lot of respect in this business, but he is also, quite frankly, an enjoyable and interesting person to spend time with. He never gives the sense he's distracted when talking with someone. He pays absolute attention to what, or whom, he is dealing with.

The shop's building is a typical, clean, industrial sort with "Roy Brizio Street Rods, Inc." and the street address hanging above a door that takes you directly into a retail sales room chock full of goodies—maybe a blown flathead or a built small-block; really, just about anything you could desire. The first person you'll probably run into is sales manager Dave Cattalini, who has been with Roy since the early days. I used to see Dave and Jim Vickery, another longtime Brizio employee, at Pleasanton during the West Coast Nationals in the early 1990s. These guys embody the friendly atmosphere encountered throughout the shop.

One of the first things you notice upon entering the actual shop area is the front-engine digger hanging high up on the wall near the entrance. Then, as your eye travels down through the shop, you realize that it's wide enough to work both sides of the space. The next area that I was drawn to was the finished-car area, complete with black-and-white checkered floor and four or five complete—or nearly complete—rides ready for test runs or pickup. When I returned to Brizio's a few months after my initial visit, a different set of completed cars was waiting in the same area. As the projects near completion, they move to the front of the building, or what might be considered the final assembly area. The rougher work, however, takes place at the rear of the building, with very ample fenced parking and a storage lot just outside the back entrance. Walking through the shop from the rear to the front, I found several projects in various states of completion. A set of clipboards helps Roy and his staff keep track of the status of projects. I watched one original-body project come in the rear door and listened in on the appraisal comments regarding the amount of work it would take to bring the body up to the standards required by Roy's shop. As the costs begin to roll up, you might question whether an aftermarket body is the answer. Of course, this decision depends on whether the car has a pedigree; if it doesn't, it becomes a question of cost or personal taste. You never know what unexpected surprises you'll find when you take the car down to bare metal.

As I continued forward through the shop, I found some stairs leading to an upper office suite from which I could look down on the floor. I did actually catch Roy in the office once—only once—sitting behind a desk. Most of the time, he was on the shop floor, elbow-deep in one of the projects or talking to customers—or even curious photographers.

One of the first things you notice after walking through the front door of the shop area is the front-engine Champion Speed Shop dragster hanging on the wall.

TEXACO
HUMBLE
DRAIN AND REFILL
Grapette
SODA
DINER
9C 76 50

A quick glimpse of John Mumford's extraordinary collection, which resides in the same building as Brizio Street Rods. One of John's most recent projects was the restoration of Barris' *The Ala Kart*, which took place at Roy Brizio Street Rods.

Ford

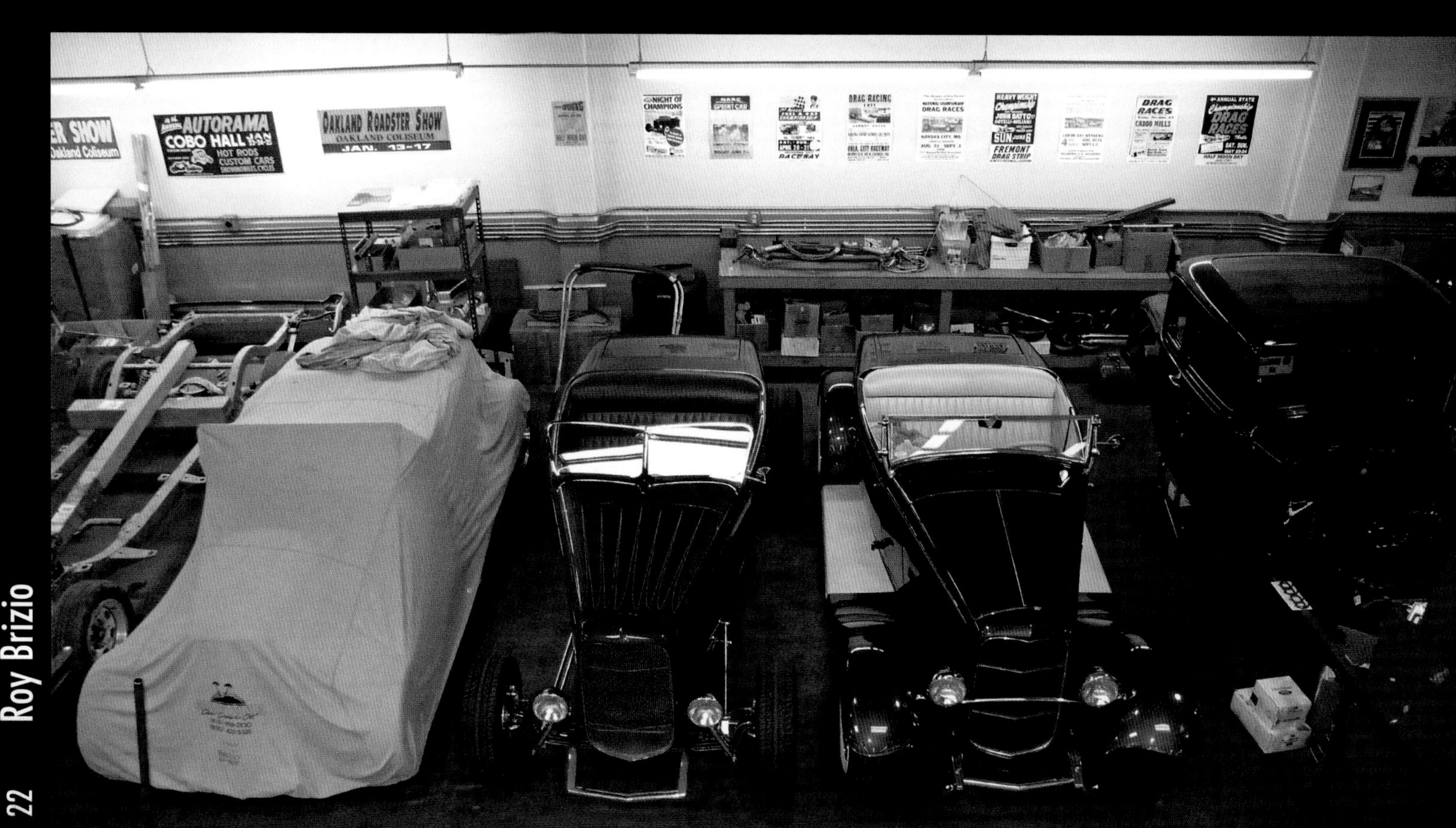
ER SHOW
Oakland Coliseum
AUTORAMA
COBO HALL
HOT RODS
CUSTOM CARS
OAKLAND ROADSTER SHOW
OAKLAND COLISEUM
JAN. 13-17
NIGHT OF CHAMPIONS
DRAG RACING
DRAG RACES
HEAVY WEIGHT
FREMONT DRAG STRIP
DRAG RACES
DRAG RACES
HALF MOON BAY

A Ferrari-powered 1933 Ford project.

n overhead storage area provides a good view of Roy's two well-known and
vell-traveled '32 Ford roadsters.

Reggie Jackson's '41 Willys.

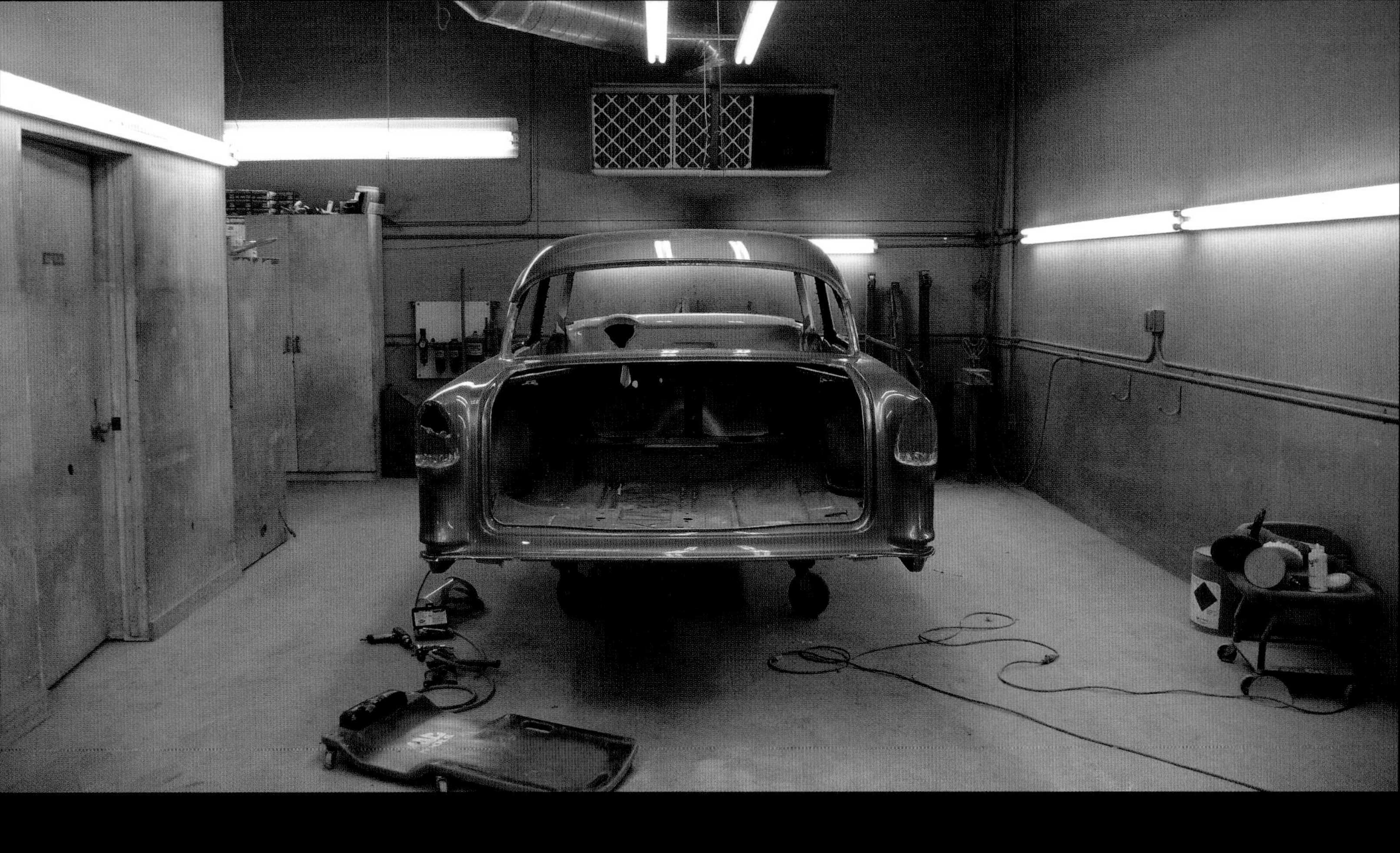

Roy's 1955 Chevy was being painted and assembled when I first visited the shop. It received the small-block shown in the close-up.

WELCOME

Looking toward the rear of the shop where the dirty work gets done (e.g. welding, grinding, and general fabrication work)

Roy's shop takes care of its customers. This '34 roadster came in for a minor adjustment.

One project at the shop when I was photographing there was this candy apple red '34 three-window owned by

Bill Ganahl at work on the Sam Barris Merc.

Paulette Zaragoza's Pagan Gold 1933 Ford Victoria is drop-dead gorgeous with its chrome wires, narrow whitewalls, and all-white interior.

The team behind Roy Brizio Street Rods.

SHOCKING!
THE DOCUMENTED TRUTH ABOUT MALE and FEMALE BEHAVIOR
VIOLATED
She danced in a way to turn even a decent man's head!
LEROY
THOMPSON

2

COLE FOSTER

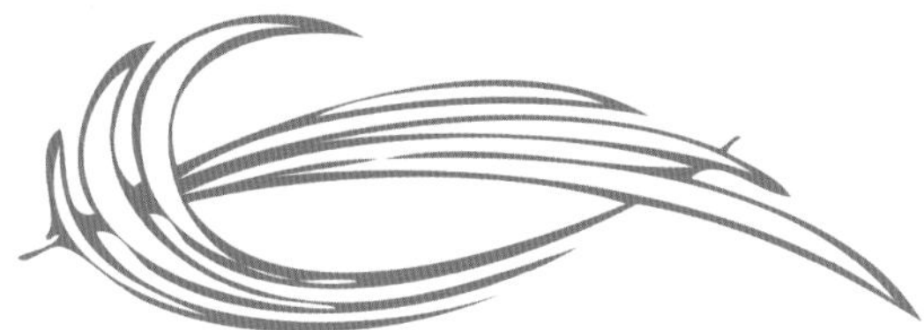

COLE FOSTER has been attracting some serious attention with his custom work on cars and bikes, winning awards and gaining recognition in both arenas. The blue '54 Chevy custom that he built is, quite simply, drop-dead gorgeous, and it proportions out as one of the best customs around. Then there is the full-custom '36 Ford that Cole and his crew built for Kirk Hammett, the guitarist for Metallica, with a full-on guitar amp built into the trunk. I first saw the car at the 2007 Grand National Roadster Show. They brought it in and parked it in the designated spot without any fanfare, turntables, or angel hair; this car did not need any of that. The '36 was parked next to Kent Kozera's '38 Ford cabriolet that had some Cole Foster touches but was originally built by Kent's father. Cole's bikes have included a number of super-clean bobbers, as well as the more recent drag bike–inspired *Moon Rocket*.

I got a chance to photograph Hammett's '36 later that spring. Cole and his wife, Susan, drove the '36 out to a lonely road on the edge of Salinas late in the afternoon. The fog was rolling in over the hills from the Monterey Bay, and it made for some interesting and moody images.

Cole recently moved out of the shop that was featured in the book *Hot Rod Kings*, by Kevin Thomson and David Perry, and I wanted to get shots of the new digs. The first thing I noticed when I found the correct address was Susan's bright red 1965 Mercury Comet Caliente out front with five-spoke Americans on it. It's the perfect accessory for the Fosters' '40s Spanish Revival house with ample shop space out back. Walk through the house or down the driveway back to the shop, and you'll find a very nicely laid-out work area, with lots of concrete, plenty of space, outside storage, and both indoor and outdoor work spaces. The shop is ample, and after some time it will probably be decorated with various photos and signage; being a new shop, such accoutrements had not had a chance to be rehung since the move. During my visit, Cole was finishing his new bike for the Paso Robles event, and I didn't get all the time I needed to photograph everything I wanted. We did get to the storage area, where Cole had stored his other well-known bikes and the '54 Chevy custom. We didn't get to talk as much as I had hoped, as there just wasn't time. Cole went back to finish the bike, so I photographed the shop some more and then headed down the road to Paso Robles.

I did get a chance to photograph Cole's latest bike at my favorite wall (great light) at Paso that Saturday before it was completely finished. The *Moon Rocket*, which won its class at the 2008 Grand National Roadster Show in L.A., now resides permanently in Japan. Next visit, maybe there will be more time available. . . .

Cole and Susan Foster, Salinas, California.

DEFEND OHIO

driveway leads back to the garage and shop area behind Cole's and Susan's house.

Moon Rocket, Paso Robles, California.

Kent Kozera's '38 Ford cabriolet also features some Cole Foster and Salinas Boys touches.

LEROY
THOMPSON

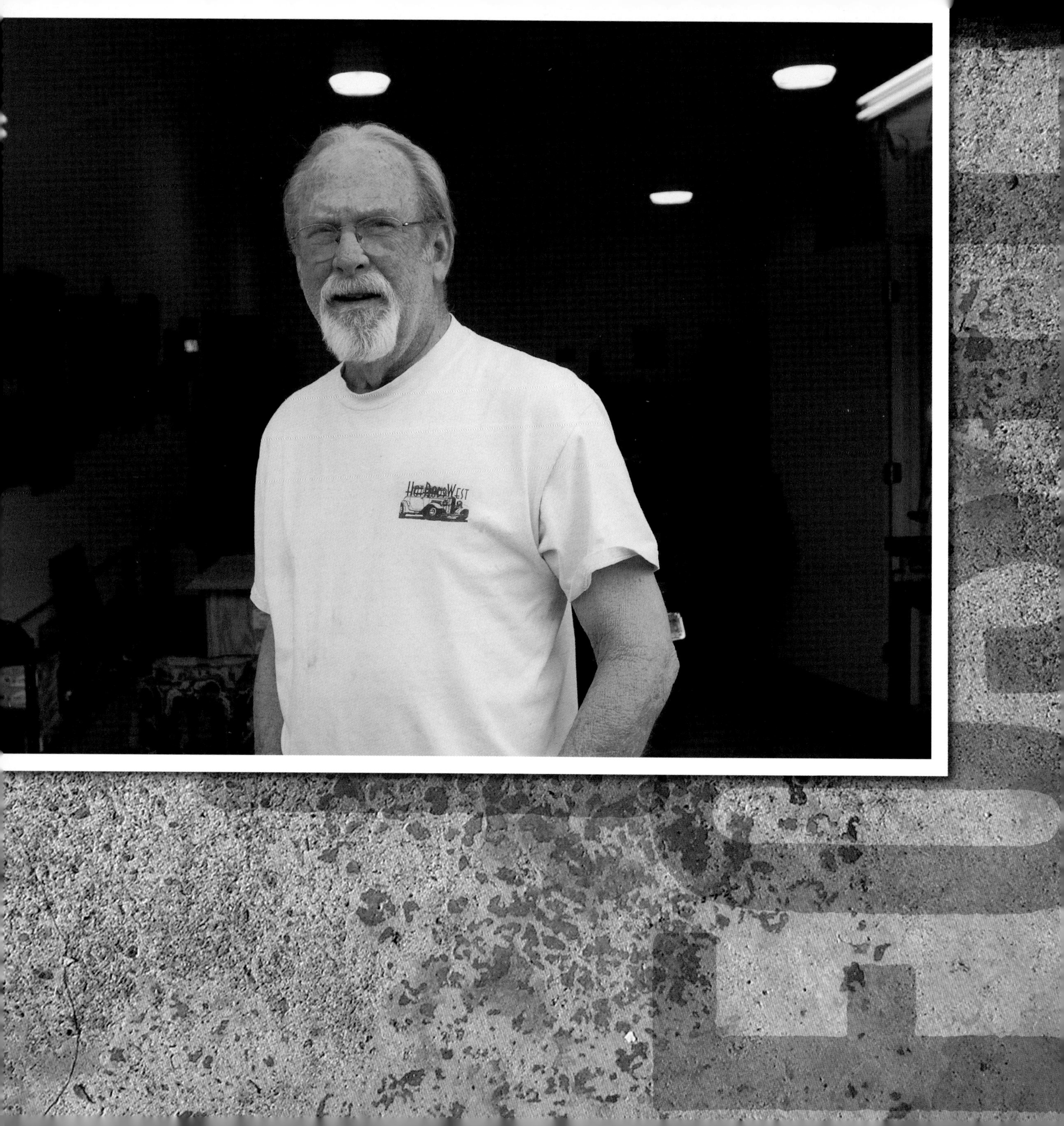
West

3
PAT FOSTER

AUTHOR'S NOTE: I had planned to get in touch with Cole Foster when I started out on this book. I had not set out to include his father, Pat Foster, until I discovered through the grapevine that he was living in Moscow, Idaho, my hometown. Moscow is a small university town surrounded by some of the richest farmland in this country. We really are off the beaten track. Nobody comes through here without trying, and it's not the first place that would come to mind as a town where you would open a shop to build and restore nostalgic dragsters. Pat did just that, however, and as it turned out, I met him before I had even talked to Cole about being in the book. After seeing Pat's work and hearing some of his history, I decided to include him. Pat put me in touch with Cole, and the result is the preceding chapter.

As I wrote this chapter, Pat Foster was in the hospital, struggling and in critical condition. I kept close tabs on Pat—and kept in close touch with his son, Cole—during that time. I have seen Cole more up here in Moscow than down in Salinas, and helped him break down Pat's shop after his illness.

I had hoped to get more stories from the golden era, but I enjoyed Pat's company too much to write everything down. Pat passed away on March 27, 2008. He was my friend, and I loved him like a brother. I really miss him, his stories, and his company. He had a great outlook on life, and he took it on face to face. He went out with a sense of grace, class, and dignity we all should hope for. Take care, Patty (Faster). I miss you.

A FRIEND OF MINE here in Moscow, Kurt Lemmon, who is a world-class custom painter, called me up one Saturday. He told me about someone I should meet who was now living and working here. I picked up Kurt, and we drove over to Pat Foster's new shop, where he was finishing an accurate re-pop of the Candies & Hughes dragster, which Kurt had just finished painting. Pat and I started talking, and a friendship began.

I had a lot of respect for Pat's history with drag racing. He was one of those who did it all, building, tuning, and above all, driving the beasts of the sport's golden era and continuing through the transition from front-engine dragsters to the first rear-engine cars. When time allowed, I spent a lot of time at Pat's shop.

Pat's new project for Billy Lynch was a re-creation of a 1968 Top Gas front-engine dragster that was originally built by Frank Huszar, Dave Jeffers, and Walt Stevens. The dragster held one record of 198 miles per hour on gas and another at 212 miles per hour on fuel. The car ran mostly in the Northeast and was well known in that area during the mid- to late '60s.

Pat's acumen and talent in putting together a front-engine dragster was a joy to watch, from the original layout of frame tubing and the front-end construction,

Pat Foster outside his shop in Moscow, Idaho.

A bird's-eye view of Pat's shop reveals an impeccably clean space—and the impeccable Candies & Hughes dragster project.

all the way to the rearend and engine installation. Perhaps most entertaining of all were his stories about why getting a set of wheels polished could be such a chore. Pat was exact about what he wanted, and he would accept nothing less. Accordingly, his shop layout was perfect and clean. This level of craftsmanship was appreciated by all who knew Pat. Notice the state-of-the-art chassis jig that Pat built and designed specifically for dragster frame construction.

Pat knew all the racers of the era and was highly respected by them. Through Pat, I bumped into Jim Hume, who had stopped over at the shop to help with some body fabrication and tank building. Jim had been working with another friend of mine, Marlo Treit, building a new streamliner for Bonneville. At the drags in Spokane, Pat and I ran into Roland Leong, whose series of *Hawaiian* Top Fuel dragsters and Funny Cars was well known all through the '60s.

Pat told me a story about when he was testing one of the first rear-engine dragsters. He made a couple of passes and said it felt good, so he put his foot fully into it on the next pass, only to leave the ground at some 240 miles per hour and hit a power pole alongside the strip, breaking the car in half (see sidebar). "Big Daddy" Don Garlits, who was just starting to run with the same configuration, called Pat in the hospital. Pat told him to run a wing to keep the thing on the ground. Garlits took the advice, stayed on the ground, and had a successful campaign.

Pat puts some finishing touches on the Candies & Hughes dragster.

Pat's shop, Pat Foster's Pro-Fab, was well known for reproduction work and restoration of some of the all-time great cars of drag racing. One of the projects he worked on was Dave West's reproduction of the Beebe & Mulligan AA/FD dragster, which debuted at the 2000 Hot Rod Reunion at Formosa Dragstrip in Bakersfield, California, sans paint and chrome. It was finished for the NHRA's Fiftieth Anniversary Nationals in 2001. Pat also worked on the restoration of the *Jade Grenade* AA/FD, which debuted at the 2001 Hot Rod Reunion, and he had a hand in the restoration of the Creitz & Donovan AA/FD.

Pat's history in this arena was immense. He drove dragsters and built them, which allowed him to understand all of the dynamics and forces at play on these machines. This rare combination of experience and ability allowed Pat to see how a car would work as it was still coming together.

Pat and I got together more than once to go cruising on motorcycles, which meant that he was constantly trying to tune the S&S carb on my stripped-down Shovelhead bobber so I could get it started. It was a kicker, and if it flooded, which it often did, I was screwed for about 20 minutes. Pat had a beautiful mid-'60s Triumph Bonneville, one of the all-time cool rides in my opinion. We had great times riding together despite the fact that each of us had to be rescued on various occasions.

This shot is from the first time I met Pat. We became immediate friends.

PAT FOSTER REMEMBERS DRIVER LELAND KOLB

I HAD THE PLEASURE of racing with Leland for three memorable weekends—weird, insane weekends, but memorable.

When builder Woody Gilmore and I decided to bring a modern rear-engine Top Fuel car to the sport in 1969, we had made a deal with "Big John" Bateman to supply one of his 392-inch engines for the car. Woody and John had some problems (imagine that!) before we debuted the new car, so Leland said we could run his 426 if we'd like. We jumped on the deal, as the 426s were coming and we knew it. Leland supplied the engine, clutch, and everything else we needed to start testing the new concept.

The first time out, we were at Orange County International Raceway and all went well until about half-track, at which time the car became very evil indeed. Woody and I talked it over and decided to park it for the night and rethink some aspects having to do with handling.

We worked the following week on slowing the steering ratio down some and went to Irwindale for more testing. Better, but still got very spooky at about 800 feet. We had brought another set of steering arms with us (just in case) and bolted those on for another attempt. Leland was calm through all this and told me to not get stupid, but just relax and let her have her head. The second attempt was again much better but still jittery at the lights. We ran right at 200 mph (clicking her off early), and the ET was fine for the speed.

During the previous two weeks, P&S made us another steering unit with a ratio of 11:1 instead of the standard 6:1 used in all front-engined cars of the era.

The following weekend, we arrived at the "Beach" full of confidence and ready to show the world the way of the future.

We had built the car with as much static weight on the rear wheels as possible and chose to not run any wing or wings until it became apparent it needed them. We prepped her for an all-out attempt (after Woody, Leland, and I deemed the handling would be sound, of course). Well, long story short, she handled like a dream, on a string, moving HARD the first half, and she settled in for a run to the eyes. About 50 feet before the first light, that bitch went straight up (first blowover?), got up on a short single wheelie wheel, launched HARD right (left tire caught the pavement first), and cleared the right-side guardrail by 5 feet! Problem was, there was a light pole just outside the guardrail. I impacted the pole at 220 something, and it tore the car off at the rear of the cage and sent Leland's engine out through the parking lot full of spectator cars and clear onto Willow Street! Oh yeah, I was still at the bottom of the light pole; those sudden stops are hell!

That ended Leland's and my racing career together but started a lifelong mutual admiration for one another.

RIP Leland. I'm tellin' ya . . . she was haulin' ass!

—Patty

BTW—So now you know how, after several phone conversations with Don Garlits and Connie Swingle (from my hospital bed), they came to design the first successful rear-engine dragster! Woody also built Duane Ong's RED that ran well before Garlits finished his.

Let's see . . . slow the steering way down and perhaps put some front wing on the bitch, and I think you'll have a starting point!

[grin]

A period-perfect re-pop of a motorsport icon. The rack at right is where Pat began every frame assembly.

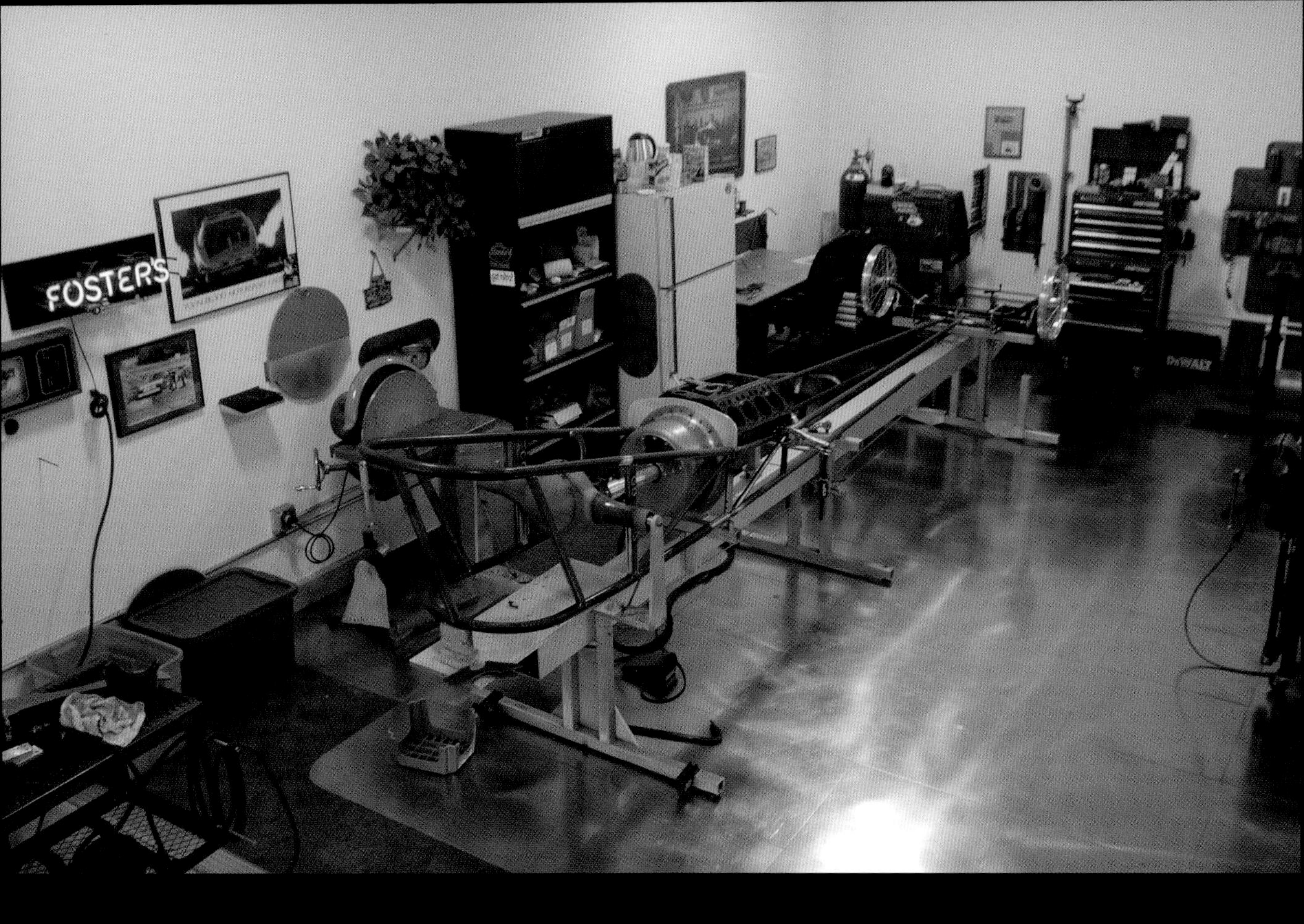

The Billy Lynch dragster project takes shape. This re-creation was based on Billy's 1968 Top Gas front-engine dragster.

Well-known and respected metalman Jim Hume and Pat Foster work out the details on the Lynch dragster project. The fuel tank was crafted by Hume and it is beautiful.

With the frame back from the powder-coater *(above)*, the final assembly of the Billy Lynch project is near.

These views of the finished dragster were taken by Lynch back East.

Pat Foster sits in his last project, the Billy Lynch dragster, just before it was shipped East.

Pat with a couple of bikes: a "project" that he picked up from his son Cole, and the mint Triumph Bonneville that Cole restored for him

DIGNEY SPEEDWAY
NOSTALGIA

4
CAM GRANT

I HAVE BEEN PHOTOGRAPHING Cam Grant's cars since sometime in the early '90s. He comes down often from Surrey, British Columbia, with cars that match the craftsmanship of any I've seen. Working out of his own garage, Cam has a unique way of putting different items together, such as placing a Pontiac hood ornament on a Chevy woody wagon, or incorporating parking/directional lights from an Indian motorcycle fender.

Cam has been building cars since he was in his teens. He got bit by the bug so badly that he's still building them with the same excitement that he had back then. He saw his first roadster when he was 12 and had his first car when he was 13—a '28 Model A roadster pickup that he got for $20. Cam found that working on it was way beyond his capabilities, so he sold it to an auto wrecker for $16. About two weeks later, he and his father found a running '31 Model A roadster for $75.

Cam says that was really the beginning. He learned how to weld with a 75-amp stick welder. He took the car apart completely and channeled it, "Because that was the thing to do." He boxed the frame with checker plate and kicked up the frame at both ends to make more room inside. He put a flathead in it for two months and then changed over to a Nailhead Buick and drove it for about six months. All in all, he spent three years or so putting together the '31 roadster.

When I ask Cam about his influences, he points to the first few hot rods he ever saw and to Roger Glassford, the fellow who bought his '31. Cam's best friend at the time was Pete Rhymer, who was way ahead of his time, marketing four-bars and other really nice stuff. Hanging around those guys really locked it in for Cam and taught him a lot.

At one point, Cam worked with Gary Lang producing DuVall windshields through a company they called "Past Tech." Cam handled the foundry and finish work before sending them out. The business was sold to "Lucky" Fred Spencer, along with one of Cam's '32 roadsters. Lucky has since sold the business. Cam says he is still involved in the production of these windshields and that it keeps him "kind of" busy. Sometime after selling the '32 roadster, Cam called and asked if I knew of any '32 three-window coupes for sale. I told him about one I saw at Bonneville, and he ended up bringing home a full-fendered, unchopped, old-time hot rod, and I don't think he really wants to change that much. We'll see if he can hold to it.

A few years ago, I flew up for the Deuce Day celebration at Victoria on Vancouver Island. I spent some time with Cam and had a chance to photograph his garage. He had just finished a phaeton project, which he left in red oxide paint. He worked so many details into the car that I found it one of his most

Cam Grant and his '39 Ford convertible, seen from his shop's office area, Surrey, British Columbia.

Some of the cars and projects in Cam's garage during this visit included his Chevy woody, a '32 highboy project with Nailhead power, and a '32 three-window.

interesting yet, and that's saying a lot because I really thought his other cars were right up there. I also shot a new project that he had started, a new '32 highboy roadster slated for a Buick Nailhead engine.

I flew up again to visit Cam in October 2007. He is very talented in many ways, and his shop is so full of artifacts from the hot rod world that I could have spent more time there just looking. He's a serious collector, but then I guess we all are in some way or another, especially if we are into building cars that are getting rarer and rarer (and the parts are getting more expensive every year).

Edelbrock
7590
7A-9392
4J 88 24
18 P 14
9D 329

TENONTHA CAMP
177 BNW

Cam's garage and attached office have a real Pacific Northwest look and feel, where the creative juices can flow.

FINALS

BELL Auto parts
WENDOVER UT
GRANT
Edelbrock
4734
D533
32 BH
43-962
18P14
9D 329
5N 398
2M3159
259 212
27 087 A
354 CGX
Ford
SALES
SERVICE

Looking out of Cam's garage into the British Columbia night. His '32 Ford roadster is on the left, and his very cool red oxide '28 Ford Phaeton is on the right. Oh, there's that airplane, too.

Cam, decked out and ready to ride in his '28 Phaeton.

Cam's 1932 Ford roadster project *(right)* is photographed next to an old Pontiac that's been there for some time. On some of his projects, Cam uses unique parts off cars like this Pontiac.

INTRUDERS

5
JOHN GUNSAULIS

I MET JOHN at one of my book signings at a well-known bookstore in Spokane, Washington, called Auntie's. After the signing, we decided to do a tour of his shop with what was left with the afternoon. I was fascinated with his current '40 Ford woody project, as well as some of the history of the shop and his father, Dick "Speedy" Gunsaulis.

Dick owned 11 '32s in just 12 years and was well known in eastern Washington state in the '60s for building show-winning cars. Dick's cars started showing up in magazines during the early '60s; a '32 Ford roadster he built was featured in the March 1963 issue of *Rod & Custom*. The car was a classic late '50s, early-'60s hot rod, with bobbed fenders in the rear, motorcycle fenders in front, and a '56 T-Bird Y-block engine with three deuces hooked to a '37 Zephyr tranny back to a '34 rearend. The windshield was chopped 3 inches, and the interior was white rolled-and-pleated Naugahyde. An article in the 1967 issue of *Car Craft* magazine that touted 1967's "10 Best Rods" credited Dick's wife, Barbara, as "a tremendous asset in helping 'Speedy' build and prepare the car to collect four hot rod sweepstakes at Portland, Seattle, and Spokane roadster shows."

John had been reworking and showing this '40 Ford woody station wagon, and I had a chance to visit his shop to get a few photographs of it that afternoon, while hanging out with a group of John's friends. John's workshop has a nice setup and an admirable collection of cars that reminded me a bit of being at Vern Tardel's shop. The space really gives a sense of how dedicated John is to the culture in general and to preserving his father's rich history and reputation for building magazine- and show-quality cars.

John seems to be heading in that same direction, as the '40 Ford has also been the subject of magazine features.

From left to right, my son Nathan, Nate Huston, Jeff Allison, John Gunsaulis, Russ Freund, and Devin Corbit hanging out in John's shop.

Gunsaulis has a nice setup and an admirable collection of ephemera that reminds me a bit of being at Vern Tardel's shop

During my visit, Gunsaulis' shop featured a '32 project and this quarter-midget racer.

A 1967 *Car Craft* featured John's father Dick "Speedy" Gunsaulis, who was well known in eastern Washington for his quality builds. That's Dick and his wife Barbara.

AUTO RACES
MIDGET
AUTO
RACES
THUNDER
BOWL
ATLAS
WIPER BLADES
Sky Chief
GASOLINE
SPEED
SHOW
Ford

6
GARY HARMS

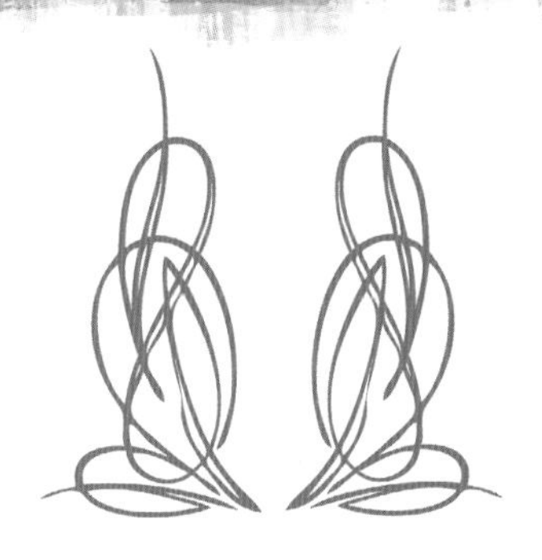

GARY HARMS HAS A HISTORY with hot rods that dates back to the 1950s, when he bought his first project, a '27 Ford roadster pickup body. He got a frame from Tim Rochlitzer, a Santa Barbara native who was attending Gonzaga University in Spokane, and through Rochlitzer, Gary met brothers Charlie and Bob Markley in 1953. The brothers were Bonneville Salt Flats racers stationed at the air force base in Spokane. The base didn't allow hot rods on it, so Gary would drive Charlie and Bob back to the base after leaving their cars in town. Gary's friendship with the Markleys eventually drew him out to the Bonneville salt.

The Markleys owed a debt to Dick Flynn, another old Bonneville racer. Flynn ran a Hemi-powered '32 coupe in the mid-1950s in which he reached a speed of 162 miles per hour. Bob and Charlie needed an engine to run in their first lakester belly tank, so they borrowed the Hemi from Dick and made some passes. To repay Dick, the Markleys eventually loaned their '29 roadster to Gary and Dick for a record try at Bonneville.

The Markleys and Harry Hoffman Sr. had built the roadster in 1979. They set a record of 210.341 in the E/FR class in the early '80s before they retired the car. Dick had been working on a de-stroked 427 SOHC engine, and Gary asked Bob and Charlie if he could borrow the beautifully built roadster to try the engine out at Bonneville. The brothers agreed, so the roadster was brought up to Gary's shop in Spokane.

Bob Markley passed away in untimely fashion before Gary and Dick had a chance to run the car, but Charlie was there while they ironed out all the kinks with the new engine. Over several years, the best speed they got out of it was 216 miles per hour. The roadster always ran straight and true and was a great foundation to try out the engine, but they ran out of patience and the roadster is now back in California with Charlie.

Gary's shop in Spokane has ample room for the pursuit of his own projects or for taking on outside projects (though I'm not sure he is really looking for any). I have seen various projects go through his shop, and Gary is meticulous with all of them, which you can see by looking at his personal cars, from the Chevy delivery to the '28 Ford roadster on the '32 frame. I'd call the roadster a keeper—a beautiful, well-thought-out, early-style hot rod that Gary did himself, with the exception of the top, interior, and chrome and metal finishing.

Gary's most recent project is a '46 Ford convertible that he and his wife, Kay, purchased as an original with mileage in around 35,000. Gary refurbished everything that was needed, realizing that this car was unique in its originality and low mileage. The car was repainted in original Moonstone Gray, the interior was replaced with a new LeBaron Bonney kit, and the top was recovered with Hartz cloth. An extra hood was punched with

Gary Harms and S.C.o.T.-blown flathead, Spokane, Washington.

Looking toward the front of Gary Harms' shop and the front end of a Chevy three-door delivery that is a great example of an early build. Gary has had this car forever.

louvers (Gary has a louver press in the shop), and the original front end was rolled out and replaced with a 4-inch dropped axle, a Durant mono-leaf spring, and a sway bar. The big addition was the 248-inch flathead with an S.C.o.T. blower and dual Stromberg 97 carburetors. Gary, it's worth noting, also has experience with wood car repair and rebuilding.

To Gary, the importance of the hot rod culture centers around self-expression. It is at its best a display of mechanical artistry that involves an individual's psyche and personality. "In the 1940s and the 1950s, when cars and parts were inexpensive or often free and readily available," he explains, "hot rodding was truly a do-it-yourself endeavor, with your personal creativity limited only by the materials and skills at hand. Today, the scene is very commercial, with a multitude of exotic and highly technical parts available. These parts are very specialized as well as costly. Entry into and self-expression through hot rodding has suffered for many because of this. Even though hot rods produced today are phenomenal, mechanical pieces of art, they are far out of reach for many. The 'rat rod' subculture is a response to this environment and may level the playing field of recognition somewhat."

Gary goes on to explain that Bonneville has such a strong appeal because it is the last bastion of truly amateur hot rodding. He says that the sport's roots in high performance are still alive and well there, more so than in other areas of hot rodding, and that Bonneville is a place where mechanical excellence is rewarded, in ways both technical and visual.

The red ’28 Ford roadster sitting on a ’32 frame is as traditional and “old time” as any
Yes, Gary owns a louver press with the old-style rounded hot rod louver dies

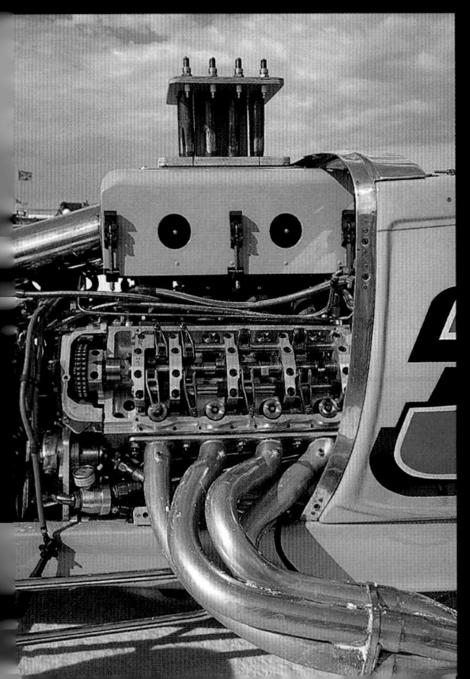

Bob and Charlie Markley's '29 roadster that Gary and Dick Flynn used as a platform for Dick's de-stroked 427. The roadster is now back in California with Charlie.

The back end of the shop, which has a hoist and, in this photo, the gray 1946 Ford convertible powered by a S.C.o.T.-blown flathead.

MOBILE
GASOLINE

7
RON JOLLIFFE

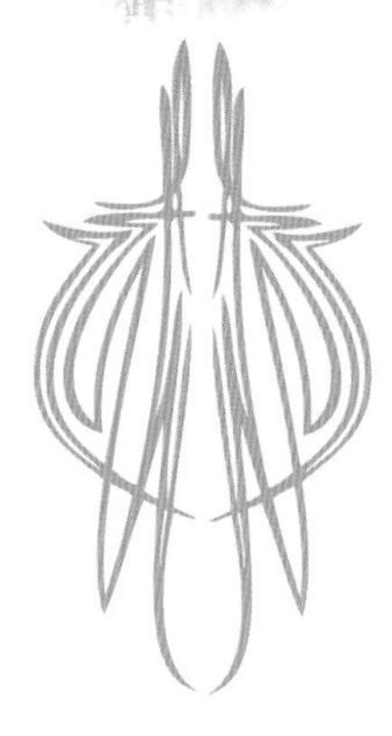

ORIGINALLY, I INTENDED to feature Ron Jolliffe's new shop in Paso Robles, California, the plans for which I've been following. Knowing Ron, it's going to be interesting and very functional. He has a knack for making things work, even when people tell him he's not following the rules as closely as they would like.

Unfortunately, photographing the new shop did not work out schedule-wise, so I decided to feature his other shop in Hailey, Idaho, which I have visited often and where I have always enjoyed spending time. I remember doing a photo session down there one winter and shooting some of it outside, as the temperature plummeted below zero and I realized that I could not feel my fingers. I have also helped crew a couple of Ron's racecars. We traveled to Muroc a few times to race his '49 Olds fastback, and I got to drive it at Bonneville (my introduction to the incurable disease of salt fever). The drive down to Muroc from anywhere in Idaho is quite a haul, but we had some interesting and fun times on the trips. Ask Ron sometime about the hourly rate Cockroach Motel in Riverside. Strange trip it was, looking at parts up in the Riverside hills that we had heard about—Indian bikes, old Triumphs, Hemi engines, Hilborn injection units, and an old Bonneville racecar, which turned into a whole other story in itself.

The shop in Hailey has produced some fine cars, two of which are Ron's racecars—the aforementioned 1949 Olds Sedanette, first running an Olds big block and then a Jimmy 300 six-cylinder (Ron likes to experiment), and a hand-built '34 street roadster that was innovative to say the least, using specially formed frame rails that dropped the car almost to the ground. You want to be low to the ground if you're running high speeds at Bonneville—air building up under the car at high speeds can lift it right off the salt. Ron also built a steel '32 Ford roadster in his shop loaded with, of course, an early, blown Olds V-8. This car scoots.

The Hailey shop is well laid out, with most of the machinery needed to build just about anything Ron wants. The high ceilings and chain-hoist system can lift almost anything. When Ron's first Street Roadster record at Bonneville ran into a challenge, he brought the '34 roadster back to the shop, hoisted it up like a carcass of beef, and cut the center out of the steel-plate floor to add a driveline loop. I still question this decision, but the bottom line is that Ron returned in October for the world finals and took the record—after he had made all of the adjustments called for in the protest complaints. Perfect.

I have seen all of Ron's cars in the shop at one point or another, as well as an early Mercedes and lots of bikes, including a Vincent Black Shadow, which Ron actually raced at Bonneville.

As for Paso Robles, I have images in front of Ron and wife Colleen's place from a photo shoot that involved Chuck Rogers' *Cop Shop Coupe*. Ron's cute little '40 Ford Standard Coupe is just behind the Rogers car. Ron purchased it from Don Small a few years back. It's in incredible shape and is the same car used in the movie *Roger Rabbit*.

Ron Jolliffe and '34 Ford roadster, Hailey, Idaho.

Jolliffe's '34 Ford roadster, on the table with its Merlin big-block engine that has run in both the A/STR and AA/STR classes.

OBILE
OLINE
LACQUER
THINNER

B/GC
1949

The Vincent Black Shadow that Jolliffe raced at Bonneville.

An outside view of the new Paso Robles shop setup.

The ’34 Ford roadster, tucked away in the Paso Robles shop...

... and in front of Ron’s all-steel, blown-Olds-powered ’32 Ford highboy.

8

"THE L. A. GROUP"

THERE IS A GROUP of guys in the Los Angeles area with whom I have become acquainted over the years. They all grew up in L.A., and some even live only a block or two from the houses in which they were raised. I have always respected their dedication to and status in the hot rod culture. Being that it is such a small world, these guys all know each other and share mutual respect and a similar approach toward whatever they build or whatever project they are involved with.

Time and space being limited, I wish I could have included many more people who are involved in L.A. car culture, such as Pete Chapouris, Jim "Jake" Jacobs, the Kennedy Brothers, Bobby Walden, Rudy Rodriquez, Tom and Diana Branch, Fabian Valdez, Danny Brewer, and Dennis Kyle. Another person I should mention is Paul Bos. I met Paul down in California, but he is now living in the northern wilds of Idaho.

Anyway, the people that I *did* get a chance to talk to this time through were Pete Eastwood, metalman-extraordinaire Terry Hegman (see also Chapter 12), Don Small, Cal Tanaka, and Billy Vinther. I have watched these guys for years, ridden with most of them, and drank a few brews with all of them. But above all, I have a deep respect for all of their work and the creativity behind their projects. How many people do you know who have the ability to fabricate any needed part? I mean really creating it out of bare stock, a sheet of flat steel or aluminum, or even the more exotic stuff like carbon fiber?

I originally grew up in the Bay Area and in southern Idaho, so how did I get so connected with all of these guys in the L.A. area? Shared interests and tastes were factors, but timing and being in a certain place at a certain time also had a lot to do with it. Dennis Kyle and I were at Bonneville in '91, and I met Paul Bos at the Goodguy Nationals in Pleasanton in 1992 or 1993. That was also the year, or close to it, that we got together with Billy Vinther, Cal Tanaka, Danny Brewer, and "Lucky" Fred Spencer (who was really from the Bay Area). After that, it was a yearly ritual to start the weekend at Tommy and Cal Tanaka's Friday-night barbecue in the RV parking area just outside the main fairgrounds. It eventually became a small community, with many of us from Southern California and more from the Bay Area.

DEUCE DAY '97
CRAFTSMAN
Roy Richter
BELL
Auto parts
KIMBAL 5728
V8
MORE
ROAD
RACE
CRAFTSMAN
ALWAYS A HIT!

GENERAL AUTO REPAIRS
GENUINE
Ford
PARTS
Eastwood
OK

PETE EASTWOOD

I FIRST MET PETE EASTWOOD sometime in the 1990s, when Billy Vinther introduced us at breakfast in a Temple City neighborhood restaurant. I had heard of him and his relationship with the Pete & Jake's shop during its early days, and I knew of his reputation as a fabricator and hot rod builder. Pete was considered a real hot rodder, maybe somewhat of a rebel, and very much an individual. "P-Wood" lives in the same neighborhood where he grew up, which is also just a couple of blocks away from Blair's Speed Shop, a place where he spent some of his formative days.

Pete does things his way, without apology. What I didn't realize, until I spent more time talking to him and other people involved in the culture, is how well known Pete is and how eclectic his overall tastes are regarding automobiles and motorcycles. His presence and history in the hot rod culture, and the Southern California automotive scene, is paramount. He has definite tastes and viewpoints when it comes to designing and building cars. I have found that his projects always are interesting and usually end up right on the money in terms of their cool factor.

One of the cars that rang my chimes was the Eastwood and Barakat red oxide painted '32 Tudor that graced the cover of *Hot Rod* magazine in November 1982. It was the first-ever primered car to do so, thanks to the influence of Gray Baskerville, the legendary hot rod journalist. That car was all hot rod, especially when taken to the strip.

Another well-known car that Eastwood had a hand in building was Tom Prufer's *Cop Shop Coupe*. The proportions on the car are extensively reworked. Chuck Rogers now owns the car, and I've photographed it a couple of times during his ownership. It has always been one of my favorites.

Pete helped build a '29 roadster project that raced at El Mirage and Bonneville with Tony Piner. The roadster ran in the 180s at El Mirage and was able to hit the high 190s at Bonneville, but it never ran over 200 while Pete and Tony partnered on the car. Pete took the weight out of the car, put slicks on it, and ran it at Fremont in the 11s to a speed of 150 miles per hour. Pete stayed with drag racing, moving from the roadster up to a dragster in the late 1980s, when he became tied back in with Blair's Speed Shop and Phil Lukens, who then owned Blair's. Pete also has been involved with nostalgic drag racing, including the restoration of the Fiat-bodied Monello & Matsubara AA/FA.

But Pete's automotive tastes transcend hot rod culture, ranging from early T speedster re-creations running bangers, which he will drive on today's modern highways, to early brass-era cars, Miller Indy racers, and traditional motorcycles and bobbers. His

Pete Eastwood, Pasadena, California.

Pete Eastwood always seems to have a 1932 Ford Tudor in the backyard. Here, you also get a glimpse of his beautiful Chevy-powered Ford 1½-ton flatbed.

combination of artistry and engineering skills makes whatever comes out of his garage a must-see. Every damn thing, whatever genre you might put it in, just looks right—a functional form of art.

The last time I stopped by Pete's house and garage, which some call "Onramp Auto Parts" because of its location near the freeway, there was a Model T truck body sitting on a '28 chassis. It was running a 327-inch small block, a Turbo 400, and a Ford 9-inch rearend. This one was finished, painted, all done. Simply beautiful.

When the creative part of a project is finished, there's always another one over the horizon, pulling Pete away from the one that's current. Interesting dilemma. To say that I have a lot of respect for Pete's creative and functional acumen is an understatement. The photo of Pete's living room should tell it all: nestled amongst the Craftsman-era furniture are a gorgeous Triumph drag bike, a brass-era radiator, a Hilborn two-port injector, an up-to-date blower and injector system, and lots of books.

I would love to see a history written about Pete Eastwood and his relationships to so many well-known—and not-so-well-known—hot rodding figures. His stories of the Pasadena and L.A. automotive and hot rod cultures from the mid-'50s through today would make one hell of a read.

You can always tell a "working" shop.

GENUINE
Mapco
DETROIT
IGNITION PARTS
STARTING
LIGHTING
IGNITION
DEPENDABLE SERVICE PARTS
SPEEDWAY COILS
Than A Good Day Working
CRAFTSMAN

Eastwood's T speedster project in progress.

Pete is into brass-era cars, too. This Stevens-Duryea is parked in one of his two garages behind his house.

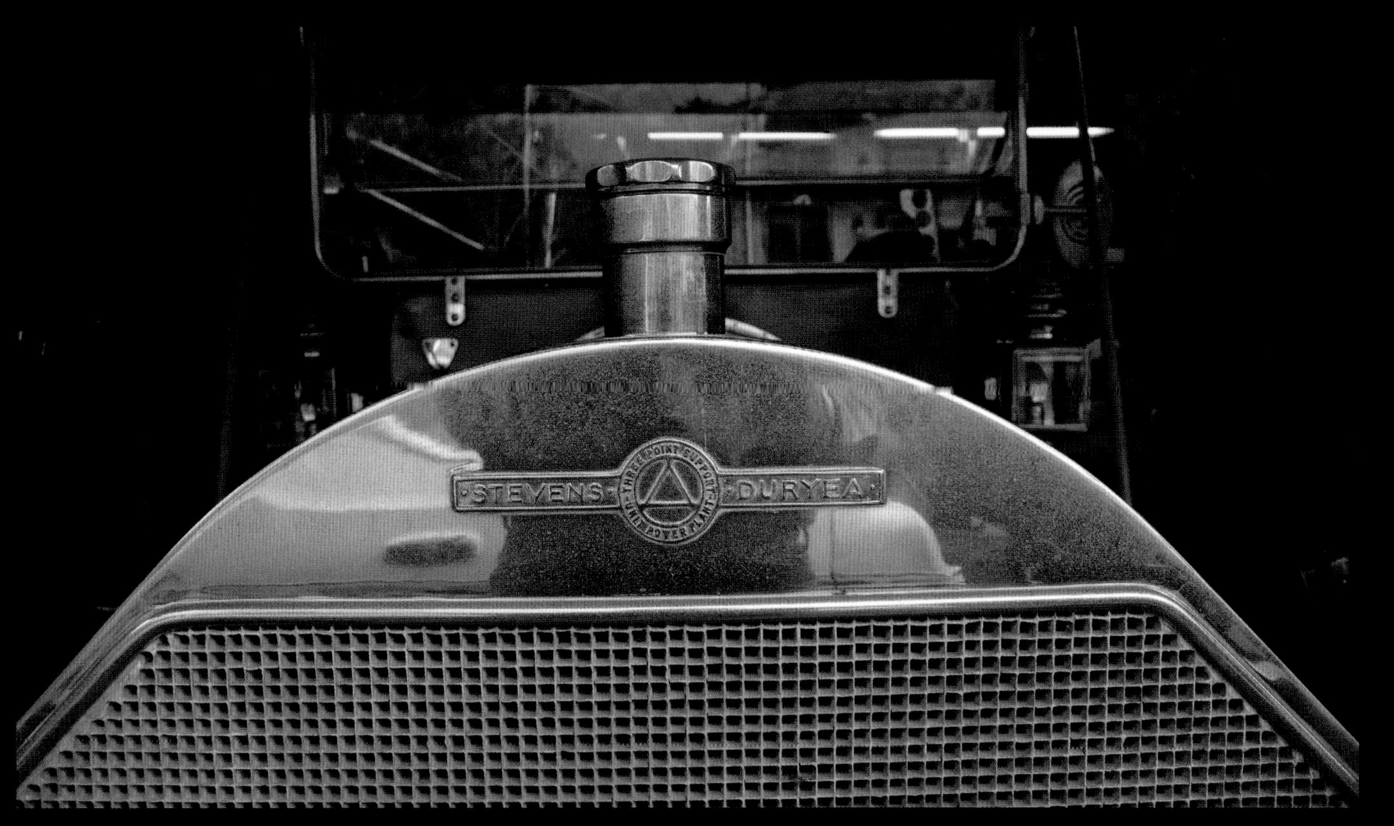
STEVENS
DURYEA
THREE POINT SUPPORT
UNIT POWER PLANT

STARTING
LIGHTING
IGNITION
DEPENDABLE SERVICE PARTS
Than A Good Day Working
Ford
The Car That Changed The World

An overall view of the working shop. "Value you can trust" is what Eastwood offers.

“Onramp Auto Parts.” The Ford truck sits on a ’28 chassis with a Model T body and runs a 327-inch small-block and a Turbo 400.

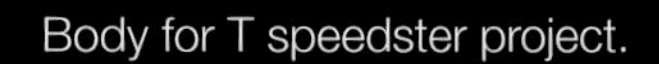

Body for T speedster project.

Here's an Altered drag race project and parts. Notice, too, the Indian motorcycle parked behind the Durye

This shot of Pete's living room should tell it all. Nestled among the Craftsman-era furniture is a gorgeous Triumph drag bike, a brass-era radiator, a Hilborn two-port injector, an up-to-date blower and injector system, and lots of books.

LINCOLN
ARC WELDER

BRICKYARD
400

TERRY HEGMAN

I WAS INTRODUCED TO TERRY Hegman through Dennis and Debbie Kyle sometime in the 1990s. They also live over on "that other side" of L.A., out toward the ocean. It all seems some distance from Temple City, Pasadena, Alhambra, Glendale, and Altadena.

Terry and Mary Hegman came to So-Cal from northern Michigan, an area of the country that lays claim to real winters. In a town called Champion, Terry started work in a body repair shop. In 1973, he opened his own shop and decided to focus on custom painting rather than normal collision work. Soon, Thom Taylor took notice of Terry's work and introduced him to Boyd Coddington. Terry started to do some paintwork for Coddington, flying out and staying until he finished whatever job he was painting.

Terry and Mary moved to Fountain Valley in 1983, deciding that Southern California was where the real work was. One such real project was Don Raible's *Blusteel* Deuce roadster, which won the 1993 America's Most Beautiful Roadster trophy. Terry had stopped painting by then, having decided in 1986 to limit his work to metal finishing. After focusing his work, he became well known for all forms of metalwork and customization. He doesn't even prime his own metalwork any more, and it doesn't need it much, which says volumes about his ability to produce finished metal that is nearly pre-paint-prep pure. This level of work takes a fine hand and a lot of talent.

I watched Terry build what I consider one of the finest '51 Merc customs ever. I've looked this car over in the pink (bare metal), and to see it all without any filler or paint was awe-inspiring. The proportioning, balance, and overall design aesthetic of this car were also astounding. There is not a bad angle anywhere. Check out the fit of the doors, the skirts, and, well, everything.

Certain creations go beyond just being good, or great, perhaps even fitting into the realm of Fine Art (with capital letters). These creations go beyond the show value of the AMBR winners. You can perceive the creator's restraint, that voice that told him not to go too far. The good ones know when to stop. Terry does, and what he left us is one of the all-time great examples of how good a '51 Mercury can be.

One of Terry's latest projects is a rework of a full-bodied, front-engine Top Fuel dragster with a body that was done originally by Tom Hanna. Terry said that he is thoroughly enjoying the project and is starting to see more and more of these revived dragsters. I also came across a gorgeous track nose that Terry had just finished for the Rolling Bones shop back in New York state, destined for their '34 Ford project for George Poteet (see Chapter 12).

Terry Hegman, Fountain Valley, California.

From the inside looking out, with Terry's sublime '51 Merc on jack stands

There simply is not a bad angle on Terry's '51. Frenched '54 Mercury headlight rings are set on both sides of the shaved and rounded hood. Terry took great pains to retain the look of an early custom inside, as well

Terry shaved the trunk of the '51 and used '54 Merc taillights as an homage to the Barris Brothers.

The Mercury grille surround houses a '54 Pontiac grille. There is an amount of restraint that tells the creator not to go too far.

BELL Special
Roy Richter
BELL
Auto parts
SPEED PARTS EXCHANGE
KIMBAL 5728
V8
CRAFTSMAN
DEUCE DAY '97

DON SMALL

"I LOVE THE HISTORY of the early days of racing in America," Don Small says, "and I find myself influenced heavily by those days and the connections that are made from there to the hot rod."

I have talked at length with Don Small about the history of auto racing in Los Angeles, especially the board-track days of the early '30s. He picked up the rights to the trademark of the old Bell Auto Parts name some time ago, so two of his cars are graced with the Bell logo. The first is his black '32 roadster highboy, the *Bell Special*, which I watched him put together, with all of the right late-1950s accoutrements. Pull off the big sugar-doughnut whitewalls, and the car could easily slip into the early '60s.

Don's latest project reflects some of the history that interests him so much—a faithful copy of a '33 Ford Gilmore Cup racer, which is essentially a stripped-down roadster. The 250-mile, closed-course Gilmore Cup was held at Mines Field, at the northeast corner of Aviation and Lennox, which is now part of Los Angeles International Airport. The American Automobile Association (yes, *that* AAA) sanctioning body dictated pretty strict adherence to stock specs, and Don has taken great pains to remain period correct. Sure, there was no No. 11 racecar at the time, but everything else on this '33 is an impeccable reflection of the era, especially the lettering, which rings absolutely true to the time. Don decided to have former Bell owner Roy Lichter's name painted on the side of the car. Lil' Louie in San Bernardino really nailed the graphics, according to Don. I agree—it's beautiful.

In addition, the detail work—the windscreen mounting setup, the sprung leather hood hold-downs, the engine-turned original dash—are also true to the original racers, which now sell in the $200,000 range. With this car, Don has honored a very short but important period of automotive racing history—one that influenced much that we see today.

"I grew up in the right part of the country," Don says simply. "The San Gabriel Valley had a good deal of hop-up and roundy-round racing activity before and after the war."

Don noted that Harry Miller, who built the most beautiful and exotic racecars of the 1920s and 1930s, had his shops in L.A., alongside the other machine shops that supported racing activity before the war. Racing was a big deal in L.A.

"I don't think hot rodding would have grown up if there hadn't been so many people interested in racing and all those clever people who supported all those racing adventures in this area," Don says. "Some of the most important racecar development in the history

Don Small with his 1932 *Bell Special* and the 1933 No. 11 Bell Ford roadster, Alhambra, California.

"I love the history of the early days of racing in America, and I find myself influenced heavily by those days and the connections that are made from there to the hot rod," says Don.

of racing occurred here in L.A. in the 1920s, and they knew how to modify their own cars by working on midgets in the 1930s. The industry that occurred here during the war taught people how to work aluminum, how to do the machine work, et cetera. And there was extra capacity and extra talent after the war. So they employed those talents in their hobbies."

Check out Don's rides, and you will find a quietness in some of the detail work and subtle little surprises. Word is that there might be a new '32 roadster in the works, with full fenders this time. I look forward to seeing it.

Don acquired the rights to the old Bell Auto Parts name some time ago. He has incorporated it into his faithful re-creations of an earlier hot rodding era. The big wide whites, stance, windshield height, Halibrand quick-change, and 327 small-block work together. A black and white T-Bird-style rolled and pleated interior adds to the late-1950s effect.

Everything is "right" on Don Small's reimagining of the Gilmore Cup No. 11 '33 roadster, from the leather hood straps to the mechanical brakes, the windscreen set up for racing (note the brass covers over the original post locations), the period-perfect lettering, and the front nerf.

Roy Richter
BELL
Auto parts
SPEED PARTS EXCHANGE
KIMBAL 5728
V8
BELL
CRAFTSMAN
DEUCE DAY '97

CAL TANAKA

CAL TANAKA STARTED his hot rod career in his dad's body shop. Tommy Tanaka was an early member of L.A.'s Outriders car club, which dates back to the early days of dry lakes racing before World War II. Both Tanakas have been influences in my life, especially when it comes to understanding life revolving around the Southern California hot rod car scene. Cal once built Tommy a '32 roadster as a son-to-father gift, a beautiful gesture for all that Tommy had imparted to Cal throughout his life.

I have photographed two of Cal's garages through the years. The first was behind Tommy's house and had its own living space attached. The second shop is Cal's own, behind a house that he purchased some time ago. He has reworked portions of it to be able to store projects just off the back of the house and still have a full two-car garage to work in. In there, he's built his own projects and done metal shaping and finish work for others. Cal has one of the nastiest, lowest '33 coupes around.

Two other projects sitting idle in one of his garages are the front-engine digger with the Terry Hegman body and the Danny Brewer–built engine, as well as a metal '32 highboy roadster project with a Billy Vinther built frame underneath. Cal's latest project is a '46 Ford coupe. He has completely freshened the frame and body, covering the latter with Tanaka black paint and Tanaka flames. The coupe was not quite ready for a road trip when I visited, so I photographed it in the driveway, next to Cal's house, which proved to be period-perfect. There was also a lowered '68 Camaro with five-spokers and original paint that I fell in love with, but which has since gone on to a new owner.

Cal Tanaka with his flamed '46 Ford coupe, Covina, California.

Looking past the Tanaka-flamed '46 Ford coupe and into Cal's current shop area, located behind his house.

Ford
AUTHORIZED
SALES AND SERV
NO FUCKIN
TRIBAL, GHOST, SEA
OR SPLASH BULLS
BEFORE FLIGHT

Cal's front-engine digger features a Terry Hegman body and a Danny Brewer–built injected small-block. It's seen in Cal's current shop *(above)*, and in the shop that he had behind his father's house *(right)*.

Budwei
BUD LIGHT
ALWAYS A HIT!

Budwei
BUD LIGHT
ALWAYS A HIT!

The working area and back of the second garage. Of course, in L.A., working outside is easy. I've even seen Cal's English wheel out in front of the garage.

A couple of cars sport the "Tanaka black and flames." The '32 in the background was a gift from Cal to his father, Tommy. His '33 coupe is one of the lowest and nastiest around. It runs a 355-inch Chevy and is completely detailed underneath with chrome and pinstriping.

BILLY VINTHER

WALK UP THE DRIVEWAY past Billy Vinther's very neat bungalow in Temple City and into his fenced backyard, and you'll find some large, clean, concrete pads leading into a garage.

The original two-car garage was built up into a viable shop that includes the original setup plus about four times the workspace. The original garage space is now carpeted and is separated from the rest of the structure, serving as a clean storage area for Billy's finished cars, including his well-known orange '34 three-window coupe and a semi-finished 1956 Chevrolet sedan delivery in the Vinther style—impeccable and low to the ground.

Just to the left of the original garage is the front entrance to the shop. That door opens into a long section that branches off to the right and leads to the back and an area that is at least two cars wide. This area is behind the original garage and has an opening off to the side. Billy could conceivably work on four automobiles in this side and back area. All in all, the shop is kept clean with everything in its place. There are myriad tools and equipment located around the shop for all the different parts of project buildups.

At one time, Billy had a '53 Olds two-door hardtop custom that he picked up from Terry Hegman and an almost-completed '32 frame set up with a Halibrand quick-change, a four-speed tranny, and a small-block Chevy with the complete front and rear suspension installed. Parked in the back area during my visit was an old semi-original '32 Ford three-window coupe with full fenders and a small-block Chevy hooked to an early original Ford drivetrain.

Billy is one of the best fabricators around, bar none, and knows all there is to know about putting together a traditional American hot rod, especially one representing the period from the late '50s through the '60s. His orange '34 has no stereo, air conditioning, or power anything. It's noisy and cackles with a high-compression, Engle-cammed 327 Chevy grumbling through the dual-header exhaust. Billy's cars are basic, but his detail work is beyond what most others would even consider, and his eye for proportions is right on. To make such cars absolutely right, and still drivable, is truly an achievement.

Billy Vinther, Temple City, California.

Billy's original two-car garage has been mostly carpeted to provide clean storage for cars and projects, including, on this visit, a Merc, Billy's well-known '34 three-window, and a '56 Chevy sedan delivery.

Behind the storage area is a larger working space. Billy could conceivably work on four automobiles in this area.

Notice the sway-bar treatment on the Halibrand rearend, the ladder bars, and the beautifully worked exhaust *(top)*, as well as the mechanical clutch and bell-crank setup *(above)*. Beautiful, adjustable, and functional.

A view from the back of Billy's shop shows his'32 Ford three-window and his '53 Olds two-door hardtop custom that was started by Terry Hegman. Both have since been sold.

This view shows the back of the shop, a frame rotisserie, and the neatness that Billy is known for.

I have photographed Billy's garage many times, and there are typically three or four rigs there. Even so, there's still plenty of space to gather and swap tall tales.

FIRE CHIEF
GASOLINE

9

BOB LICK

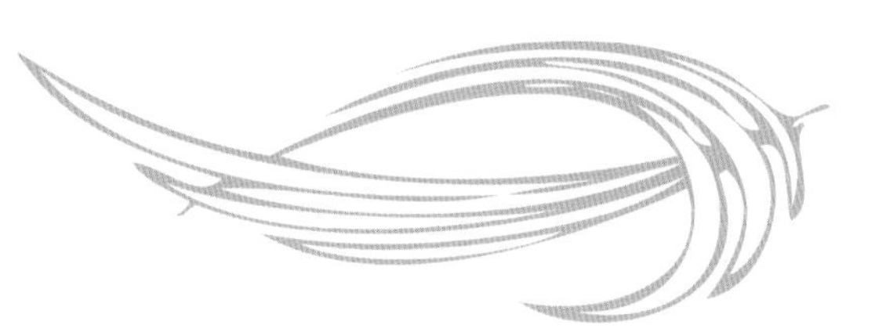
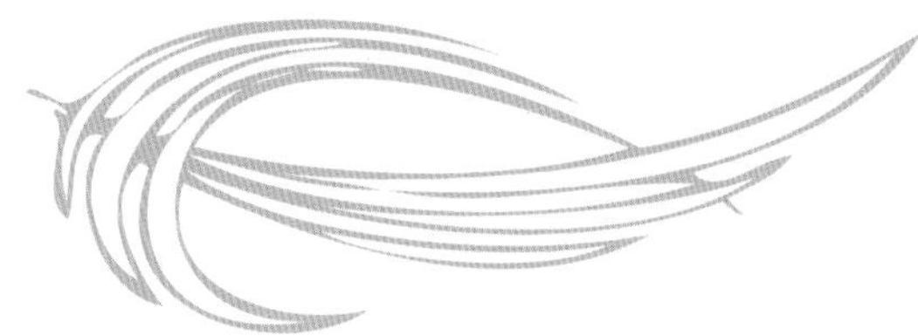

I HAVE KNOWN BOB LICK since the late '80s or early '90s. I started photographing his cars when he began showing up at events with this black '32 Ford three-window coupe that was full-fendered, low, subtle, and flathead-powered. He came out of Baker City, Oregon, so we set up a photograph session there in 1994 with Bob and a yellow slammed '40 Ford Standard Coupe that he was starting to rework, ever so slowly, into a drag racer, constantly getting more and more serious about the racing side of the hot rod equation. Bob's old black '32 arrived for part of the shoot, and Jim Lindsey brought his '32 Ford roadster highboy over to be part of the session. Dale Withers drove over from the Portland area (Eagle) with his black and oh-so-nice '40 Ford Deluxe Coupe with a Columbia rearend. We spent the day, the evening, and part of the next day photographing these cars. Bob's '40 had no headlights wired in yet. He was racing it and didn't really need them, but he realized partway into the sweet light of that first evening he had to scoot back to the shop before it became completely dark.

Bob was a big help to me when I was building my '48 Ford woodie, since sold. As it turns out, I now have a '40 coupe, and Bob has come through with some desperately needed parts.

I took another trip down to Baker City to photograph Bob's coupe and to get a few shots of his shop for this book. Bob had kept on racing the coupe and improving it since I first saw it in 1994. He changed the front-end setup and installed a serious big-block. The car now sits on a stock '40 frame and is back on the street with a stock Ford drivetrain hooked up to a '52 324-inch Olds engine with three deuces. The car seems to sit even lower than it did when I photographed it in '94. The primer spots on the passenger side and rear are the result of getting the car backward on the strip once.

Forty Fords were *the* car in my day, absolutely equaling '32s in popularity. I would have trouble even today making a choice between the two. On the shoot for this book, we took Bob's '40 out for a new maiden voyage, as Bob had just gotten everything put back together. He let me take it for a spin out on the highway. It handled beautifully, and it was as cool as it could be.

Bob's shop is in the same place it was when I first met him, and it's definitely dog-friendly, which is part of the scene down there. The shop looks pretty much the same as it did in '94, with the parts and projects representing the main changes. At the back of the shop is a beige Ford pickup that's going to be just a little different. Bob's not going to change the body or paint in any way, but the suspension and drivetrain are race-bred. Bob inherited a complete race chassis, and he is working it into the truck. This pickup will be the full deal, with race hubs, a beefy quick-change rearend, and the struts and four-bar system to hold it in place.

The one event these Oregon guys never seem to miss is the Bakersfield Hot Rod Reunion in the fall. A couple of them also are starting to show up at Bonneville, and Bob keeps saying that he's going to make it out to Speed Week sometime.

Bob Lick and his 1940 Standard Coupe, Baker City, Oregon.

Bob's taste leans toward the traditional and he built his '40 with nostalgic drag racing in mind. When I visited him in 1994, the coupe had a lumpy 383 Chevy small-block, but the stock Ford drivetrain is now hooked up to a '52 324-inch Olds topped with three deuces.

Bob's shop is pretty much the same as when I first visited in '94.
Bob raced his '40 across the West for some 10 years without ever taking it off the original frame.

The yellow paint on the '40 has remained the same. It's a classic revival and perfectly fitting an early vintage hot rod. The primer spots on the front and rear passenger side resulted from getting the car backward on the strip once.

Both of these photographs were taken in the 1990s and feature Bob's previously owned, full-fendered '32 Ford two-window and Dale Withers' since-sold black '40 Ford Deluxe Coupe. The '32 roadster behind Dale's '40 still belongs to Jim Lindsey.

At the back of his shop, Bob has a Ford pickup which he says is going to be "just a little different."

10

STEVE MOAL

AUTHOR'S NOTE: My first introduction to Moal Coachworks was out on the return road to the Bonneville Salt Flats. Phil Linhares introduced me to Ed Adams of Berkeley, California, who was driving a hand-formed, aluminum-bodied, beautifully designed car that was sitting on a Triumph TR2 chassis. As it turns out, the car was built by Jimmy Kilroy, a metalman who works for Steve Moal at Moal Coachworks in Oakland, California.

The second time that I was introduced to Moal Coachworks was again with Phil Linhares, who took me by the coachworks to see a new project that Steve Moal had in there. It was in the building that Jimmy Kilroy works in, which is across the street from the Moal Coachworks main building. Jimmy was working on Eric Zausner's hand-built V-12 Ferrari Marinello–powered roadster. Hindsight tells me that I screwed up by not getting some photographs of the car at that stage of its construction.

Steve runs a shop that is renowned for a very high level of craft and creativity. I would have to say that some of the people working for Moal Coachworks are artisans and craftsmen of the highest degree.

Steve is very conscious of the design process and of the fit of each project to each owner. These are true coachwork approaches, and as we were talking and going over some of his drawings and conceptualizations, I became fascinated with the complexity of the projects and of all the process- and detail-oriented thought that goes into each build. The drawings on the design room wall were gorgeous. I found myself taken back to when I was a child and automobiles were truly designed for their aesthetics with, of course, some consideration to functionality.

Steve took me over to the shop where Kilroy works and introduced us. Meeting Jimmy Kilroy at Steve Moal's shop some years after that photo session at Bonneville kind of closed the circle. I spent a good deal of time just observing the patience that goes into Jimmy's work and admiring some of the finished pieces that were around the shop. Steve also showed me a platform that a client sat on in order to properly locate everything that is pertinent to the driver's personal comfort and the car's drivability.

Steve Moal, Oakland, California.

To say Steve Moal's home garage is well-equipped would be an understatement.

A view through the shop from the front entrance, with Hank Torian's '46 Moal Chrysler *Aghassi Royale* in the foreground, and Tim Allen's '55 Ford *Triple Nickel* project beyond. Notice the second-floor office space in the upper right.

The House of Moal

By Michael Dobrin

The portal to a muted green building on Oakland's East 12th Street is secured inside by a simple steel bar. Pivoting on a center pin, the black bar, about 3 feet long, has forged protrusions at opposite ends that serve as handles. In locking mode, the ends of the bar swivel into opposite-facing, up-and-down, steel channels. Utilitarian and strong, that bar is, in Steve Moal's words, "A symbol of permanence."

It was fired, pounded, and shaped, probably in just an afternoon, by Moal's grandfather, William, who used it to secure the paint-bake oven at his Auto Metal Works shop in Oakland, possibly in the 1920s.

Permanence *is* the hallmark of the Moal story. It is a sense of permanence derived from trust in one's hands, from the knowledge that, with time and sensitivity to the sciences of engineering and metallurgy—and, yes, modern computer, wiring, and engine technologies—the hands can shape and create these elegant motoring conveyances, rakish and daring machines that appear to cheat the wind—even at rest.

And this building—purpose-built in 1946 to house the body and fender repair business of Bill Moal and Sons—is the perfect house for these unique works

This early Ferrari has received some of Jimmy Kilroy's special metalworking care.

to take shape and form. Within its rather cavernous 10,000 square feet, there is an immediate feeling of handiwork in progress, things getting done in metal by men bearing hand tools, arc welding and hammering, sanding, grinding—the staccato symphony of the step-by-step assembly of vehicles lined up on either side of the building.

It's the diversity of rolling stock on display that might provide the first clue that this is indeed much more than a classic hot rod shop. Yes, we can immediately identify the classic lines of a '32 Ford coupe, but this Deuce has the exposed chassis components of a high-performance sports machine; there's a sleek black Ferrari 250 California roadster undergoing full restoration; a monstrous 1932 Lincoln roadster, now in a state of deterioration, awaits its rebirth months down the road. Big Westlake Chevy and Ford Cleveland engines sit in various coupes and roadsters. In the paint booth, a craftsman lays the first of many coats on a beautifully shaped pontoon fender.

Diffused light from ample skylights flattens and softens the hues of the metal and paint throughout the interior. The light gives each project a patina in

Dennis Varni's new 1933 Ford roadster project.

Tim Allen's 1955 Ford, *Triple Nickel*, runs a 5.4-liter DOHC Ford GT engine and a Ford Lightning tranny.

black, white, and gray, much like a photograph taken early in the last century of an American highway garage or an Italian *carrozzeria*. The arched wooden ceiling beams are braced by red steel trusses. The walls are a simple, atonal brick.

From his lofted office, the 62-year-old Moal points to what is not immediately obvious—and to some of the reasons this particular garage supports the continuity of motoring arts under the Moal signature.

"My father was a machinist in the San Francisco shipyards during the war, and he and his dad wanted their own shop," Moal says.

Swiveling to look over the floor, he continues, "You'll notice the building is entirely free-span. There are no posts. The floor space is clear. You can drive a tractor-trailer rig right through the center of the shop.

"They put up the steel frame first, then the roof, and the walls went up last. Steel pillars are capped by those brick walls, and those exposed steel X-braces squared up the building when there were no walls. There were two houses here in 1946, and the building was built on lots that were larger than the standard 50 feet; they were 65 feet each, so the building is wider. We can work on both sides of the floor.

Steve's office is the business center of Moal Coachworks
Steve's wife, Teresa, is often there making sure that all of the books and projects balance

This view from the rear of the main shop shows the paint booth on the right and a Ferrari that was in for some freshening up.

"The builder was Abe Doty of Oakland's Ed Doty and Sons. He was a big industrial builder around here. My grandfather and father must have seen other buildings like this in the area. In fact, one guy who wanted to either buy or rent this building when it went up was Qvale," Moal says. Kjell Qvale, now 88, is the Norse-born founder of the British Motors Corporation dynasty on the West Coast. In 1946, he was just getting started with a Willys dealership in nearby Alameda.

The Moals are artisans and craftsmen—whether working in wood, steel, glass, stone, or leather—they can take a long look back in history and savor what is treasured today for its singular beauty.

"There are wealthy people that can afford to invest in something truly unique," Moal explains. "They value the tradition of hand-built items—and in our case, that just happens to be hand-built high-performance cars."

The sense of permanence that imbues every inch of this building flows back in time, before William Moal emigrated from France to Oakland and set up shop in 1911. In the late nineteenth century, the senior Moal apprenticed as a wheelwright in his native Normandy.

"He learned to make wooden wheels for a carriage maker in Brest. He probably had to live on the grounds and pay for his room and board," Moal says.

Apprenticeship was a rigid and demanding practice in which the novice might spend up to five years mastering a specific craft. Moal's craft was directly linked to the guild system, the economic engine that brought untold wealth to merchants, kingdoms, and duchies in Medieval Europe. Craft guilds were also a unifying force that ensured, among other considerations, that the products created for extensive trade and domestic use were of the highest quality.

Several early mahogany-toned wooden Moal wheels—simple, functional, and strong—grace the office walls of the Moal shops, each providing a symbolic and tangible link between our digital world and the hand-craftsmanship of the Old World guild system.

But it was not the late nineteenth century Industrial Revolution or the lure of jobs in America that brought William Moal from Brest to Oakland. It was love—a chance meeting in Paris with Marie Rose Prenveille, a beautiful University of California student from Berkeley who was on a French language sabbatical.

He followed her west to the Pacific. The passenger manifest for the *LaTouraine* departing France for Oakland, California, on October 24, 1910, includes one Guillaume Moal and lists his occupation as "coach builder."

They married in California, and he set up shop on Oakland's Broadway. Even today, with its auto dealerships, repair shops, and upholstery and auto electronics garages, Broadway reveals vestiges of its earlier importance as a center of automotive activity in the East Bay. The Moal heritage, in fact, is both product and reflection of Oakland's history and geography.

In 1869, Oakland was the western terminus of the Transcontinental railway. Concurrent development of the Oakland Estuary (sometimes known as Oakland Creek and a favorite haunt of young Jack London in his oyster pirate days) as a world shipping port spawned a century-long industrial expansion in the East Bay. Foundries, canneries, shipbuilders and shipfitters, machine shops, drayage enterprises, jute mills, breweries, flour mills, pottery and tile enterprises, lumber mills, ore crushers, and farm produce brokers and shippers flourished on the east side of the San Francisco Bay.

By 1880, Oakland's population hit 35,000, just about a quarter of those inhabitants in the shining city across the bay, but many of Oakland's settlers were craftsmen who brought special skills to keep the wheels of industry turning in the booming community. The railyards alone brought welders, steamfitters, and mechanics to the area. Given this predilection and precedent, it is no surprise that the motor car found early acceptance here.

Alameda County supervisors began a road-building program in the late 1890s, one reaching out to the then-agricultural eastern portion of the county. In 1906, Oakland was the first city in the west to use a police patrol car.

Access to rail distribution and growing commerce drew William Durant and his Chevrolet Motor Company to Oakland, and his Chevrolet assembly plant opened on a 7-acre site in the Elmhurst District in 1916. In the '20s, chamber of commerce boosters proclaimed Oakland as the "Detroit of the West," and other manufacturers followed to the region. Fageol Motors built trucks as did Peterbilt, and there were Durant, Flint, Star, Fisher Body, and Caterpillar plants. Chrysler built Dodges in San Leandro, and Ford created a huge assembly and distribution center in nearby Richmond. There were engine manufacturers such as Hall-Scott in Berkeley.

The Oakland City Directory for 1915 lists 35 garage and automotive repair facilities. One among those was the firm of Moal and Fedeerle Auto Metal Works at 2935 Broadway. William Moal quite handily applied his finely honed skills to the motor trade—there was lots of work.

"He could do it all. He repaired bodies, crunched fenders, and busted radiators, which in those days were soldered up with lead. He and his partners, of course, repaired wooden spoke wheels, rims, and the like," Moal says.

"He was an extremely good welder. I remember him shaping header pipes. He was beginning to make bodies for early racecars."

In the early '20s, William Moal created coachbuilt bodies for three daring and *avant garde* roadsters: the *Battistini Specials*. Named after Mondo Battistini, a San Francisco–born newspaper entrepreneur (he sold subscriptions to the Italian language *Il Sole* newspaper to Italian grape farmers near his valley home in Stockton), the sleek, razor-fendered roadsters rode on contemporary 124-inch Buick chassis and were powered by Buick's 60-horsepower six-cylinder engines.

Moal's handiwork is evident in the steel panels, cone-shaped hood and cowl, perfect recesses for twin side mounts, and pressed-in moldings surrounding the cockpit. Two Battistini roadsters survive, one in Northern California and one in Australia.

A photo of that first shop is mounted within a small glass showcase on the second-floor landing of the current Moal enterprise. In it, front and center, the handsome, burly William Moal holds an acetylene torch. He's eyeing the camera, goggles cocked back, a bit of a sly grin beneath his drooping moustache. The torch is in an exaggerated position, almost like a conductor's wand. A number of shop mates—one of them probably his partner—are at his sides, looking a bit dour and serious.

From what we might discern, he's ready to draw a bead on a radiator core. His heat shield is but a stacked, U-shaped pile of bricks on a work table. All about the shop are various indicators of chaos and grit. One might imagine acrid odors bubbling from annealing vats and bubbling lead pots, mixing with the distinctive aromas of motor oil, gasoline, chassis grease, solvents, and spent welding gases. The aural senses would no doubt be assaulted by peen hammers, the dull thumps of shot bags against bare metal, and grinding wheels reducing hard metal in a shower of sparks—the whole effort dedicated to pounding, crimping, bending, and shaping forms both functional and decorative.

Moal and his partner applied skills that seemingly defied the mass-production method behind most automotive production, even then. They created utilitarian objects with elegant design: flared, dimpled, and raised fenders, beautifully varnished spoke wheels, riveted body parts, louvered hoods, and metal latches.

As the automobile became an integral element in not only the East Bay but throughout California, the Moal works prospered and grew, as did the Moal family. The union produced nine children. Sons George, Rene, Don, and Ben all gravitated to automobile service businesses. Ben was an upholsterer. Rene, who would later have a direct influence on Steve's workmanship as his employer, was a body-fender metal specialist, painter, and boulevard hot rodder. Don was a body-fender man and sculptor.

"George learned his trade by being the son of a specialist and, in a sense, carried on the apprentice ethic into the body and fender business," Moal says of his father. "They worked so hard in those days. My dad would put in a full day, come home, have dinner, and go back to the shop.

"My dad liked Duesenbergs and big roadsters from the '20s and '30s," Steve continues. "He said that the '32 Ford was a poor man's Duesenberg.

"While my dad liked his cars, he was also crazy about boats," Steve recounts. "He hung out with Lon Gradetti and George Mateucchi, and regularly visited his buddy Don Philbrick at his boatyard just a short drive away on the Estuary. Gradetti built and raced his *California Kid* hydroplane in the '50s at places like Clear Lake and on Oakland's Lake Merritt for the big Fourth of July speedboat regattas. My dad shaped cowlings and aluminum coverings for Lon's speedboats . . . and beautiful one-eighth scale models of Philbrick runabouts. We still have those here today," Steve says.

Leaning back toward a low bookcase, he pulls out a frayed scrapbook. Gingerly leafing through the brittle pages, he turns page upon page of cutouts from speedboat magazines of the '30s.

"My dad gave this to his buddy Philbrick. They went to high school together; in fact, we all went to the same school—Oakland High—and 60 years later he was down at Philbrick's boat shop and asked, 'What the hell ever happened to that scrapbook I gave you years ago?'"

Philbrick, without missing a beat, replied, "Right there on the shelf where you left it."

As Steve's coachwork commissions flourished, George Moal brought his considerable craftsmanship to such fine touches as hand-laminated wooden steering wheels, such as that seen on Tim Allen's *Licorice Streak Special*, and unique dash gauges that integrated harmonious colors, big numbers, and elegant dials.

George Moal died in 2002, but his legacy and influence were passed on through the craftsman ethic: learn what your father does, and make that your life's work.

"Just like my dad, I learned by watching," Steve says. Moal married Theresa Avila in 1965. Sons Michael and David have integral roles in continuing this coachwork heritage. Michael is a highly skilled body man, but a great deal of his time is spent on

Steve's grandfather, William Moal, (holding torch).

Michael Moal, project manager, in the main-floor office.

project management. David has an engineering degree and directs the computerized, technical side of the business. He has a keen eye for the fit and fabrication of components.

The Moal family home was just off Park Boulevard, a '20s-style divided avenue that skirted the upscale enclave of Piedmont. After finishing school and his *Oakland Tribune* newspaper route, Steve would ride his bike down to the Moal shop and start fiddling with stuff.

"My dad taught me how to weld, and I worked on body and fender jobs," Steve says. "I just couldn't get enough of this place. This is where I preferred to spend my time.

"I was a car nut from the get-go. Just down the street was the Circle Drive-In, and the custom car guys and hot rodders would hang out there. My uncle Rene was into the hot rods, and he and my dad were friends with Tommy the Greek. He was my hero because he could customize a car with his paintbrush."

Tommy Hrones, the self-proclaimed "Sultan of Stripe," is integral to the American hot rod story. The son of Greek immigrants, he learned his trade from an unsung Oakland stylist of his time, Niels Hoegsberg.

Hoegsberg was an itinerant sign painter who went from shop to shop doing touch-up work in the '20s and '30s, applying his skills by retracing accent lines so popular on early American vehicles. Hoegsberg also did gold-leaf applications and baroque striping on local fire engines, the most dramatic example of which—the 1906 San Francisco fire and earthquake Oakland city pumper—is on display in The Oakland Museum of California.

Flamboyant, impish, and feisty, Hrones elevated automotive painting and striping to an art form during the heyday of the West Coast hot rod and custom car movement of the '50s and '60s. His signature teardrop flourishes marked a car as having been painted by the Greek. And he was known to say, with his own bravado, "If your car wasn't done by me, you wasn't nothin'."

In the postwar years, California was a boiling Petri dish of motor racing—everywhere, all the time. The motor-crazy Moals didn't miss out on the action.

"We saw the indoor midgets at the nearby Exposition Building (site of the original Grand National Roadster Show) and went down to Hanford Speedway to see Indy cars and sprint cars." Steve says.

"We had all kinds of drag strips here—Fremont (Baylands Raceway), Half Moon Bay, Kingdon. We were doing bodywork for Cecil Yother before he drove the *Melrose Missile*. We did work for Dick Beith on his Bonneville and Indy cars.

"And of course, we went to the shows. I must have gone to 50 shows in the mid-'50s. The greatest of them all was down the street—the Oakland Roadster Show. I'll never forget walking in the door, and there was Norm Grabowski's flamed T with that tall shifter with the skull on top. That was as wild as it could get. The Kopper Kart. Ala Kart. Mantaray—saw Jeffries drive it in. Ed Roth was painting T-shirts. Saw Tommy the Greek striping Lon Gradetti's *California Kid* hydroplane. The Roadster Show was eight days long. A big deal."

With his paper route cash, Steve bought his first car, a '53 Dodge.

"It had a little Red Ram Hemi in it, so that was a pretty good thing to start with. I chromed the valve covers and radiused the wheel wells. After I sold the Dodge, I became a little more serious and did a really nice '57 Ranchero, which I exhibited at the '66 Oakland show—the last year in the old Exposition Building.

"I always had Chevelles and Dodge Chargers with super lacquer paint jobs," Steve says. "They were clean and flawless, lowered and painted up real nice. Heck, I even washed my cars every day. My dad thought I was nuts.

"In my high school yearbook, I wrote that I wanted to be custom car builder. Now in those days everyone wanted to be a George Barris, a Joe Bailon, a Gene Winfield. Doing chopped Mercs with Hall tops. That's a tough way to make a living, so I joined my family in the body and fender business. Actually, understanding the collision repair trade was important. A lot of custom car guys never learned the fundamentals of bodywork, how a vehicle is structured.

"In 1971, I took over the shop here, and we became Mercedes-Benz specialists, which allowed us to commercially form a high-end niche. Mercedes owners were very particular. We also got to work on lots of special vehicles like 300 SLs.

"I continued to build hot rods as my personal rides. I did a T-bucket and finished my '32 roadster, *Nut'n New*, in 1981."

The neat and conservative full-fendered classic Deuce was, in Steve's words, "too smooth and predictable." He was already looking beyond such styles.

The flawless aluminum bodywork of local metalsmith Jack Hageman made an impression. Hageman created everything from dragster bodies to sports car specials (the Jaguar-powered *Hageman Special* and the Chrysler Hemi-powered *Barneson Special*).

"A lot of guys wanted to do this work, to apprentice under him, but I'm sure he had little time to teach and train someone," Steve says. "You just had to figure it out for yourself. The only thing I knew for sure was that there was only a handful of people who could do this work, and I wanted to be one of them."

More diverse projects came to Moal in the '80s. He rebuilt the sleek belly tank body for the Varni and Barnett Bonneville lakester, which was a restored version of the famed Markley brothers salt flats record holder from the '50s. Dennis Varni commissioned Moal to create a special custom-fitted aluminum and leather luggage rack on his red '29 A roadster, winner of the America's Most Beautiful Roadster at the '92 Oakland show.

Moal did the metal and paint work on an immaculate '34 Ford phaeton owned by fellow hot

The front entrance off of East 12th Street, with a glimpse of Hank Torian's boattail *Aghassi Royale* and George Poteet's bare-metal '32 Ford five-window project.

rodder Tom Walsh. The car was a *Hot Rod* magazine cover piece in April 1990.

Moal created a new grille, body, and windscreen cowl on an early '30s Miller track racer for Northern California collector Ed Hegarty and reconfigured a classic sprinter that was driven by the late Pat O'Conner. He made components large and small for a stream of local hot rodders—steering wheels, hoods, fuel tanks, grille shells, belly pans, and interior panels.

There were commissions from the Pebble Beach crowd for restoration, re-creation, and replacement projects for high-dollar Concours d'Elegance show cars. The Moal bodywork on Bob Lee's 1938 Alfa Romeo 8C2900 B Spider, with its distinctive chromed brass rear wheel vent strips, helped that car take first in class at the 1990 showing. Moal's metalwork on John Mozart's 1938 Type 57SC Bugatti was stunning; the elegant roadster won Best in Show in 1998.

Along the way, Moal developed a friendship and collaboration with Hoosier 100 sprint car champion Jackie Howerton. The men developed experimental projects that integrated traditional, no-frills, dirt-track styling and track-T beehive grilles, Halibrand wheels,

Steve Moal's drafting and design area just off the second-floor office area. The drawings on the wall detail an upcoming project.

and big, exposed filler caps into high-performance street rods. Steve has no difficulty melding these enduring forms from Midwest dirt-track racing into his coachbuilt creations.

Steve's enthusiasm for the eclectic has allowed him to break free from the predictable and repetitive. There is evidence of those eclectic influences in every corner of the Moal office loft. The walls are replete with richly detailed color prints reflecting a world of automobilia. There are color renderings of projects past, present, and future; these highly changeable cartoons or portrayals of works to come are like Post-It process notes, revealing ideas and changes along the way.

There are Stanley Wanlass bronzes of stately, long-nosed roadsters in motion—the wind streaming a beautiful woman's hair while her passenger, a speed-lovin' canine, has a direct bead on the fast way forward. The collection also includes smooth, rounded, and highly streamlined stone car sculptures by Don Moal. The bookshelves are crammed with illustrated volumes on the greatest marques and coachwork houses in history: Ferrari, Alfa Romeo, Mercedes, Touring, Zagato, Pininfarina,

Steve Moal's background, energy, and vision were solidified in the early 1990s with the addition of metalsmith Jimmy Kilroy.

Harley-Davidson, Maserati, Jaguar, Cunningham, Ford, Auto Union, Scarab, Chaparral, DeTomaso, Packard, Cadillac, and Duesenberg. There are tomes celebrating board track racing, drag racing, and famous racing venues, such as Indianapolis, Bonneville, Le Mans, and Monza.

About 15 years ago, all of this background, energy, and vision finally solidified with the addition of metalsmith Jimmy Kilroy to Moal's team. The quiet, patient, and exceedingly talented craftsman sculpted coaches to fit the dreams and desires of Moal's patrons and customers.

"I met Jimmy through Patrick Otis when were doing a Ferrari restoration," Steve recalls. "We'd taken on a Fiat Zagato double-bubble, and I told Patrick that I desperately needed some help. He'd just sponsored Jimmy, who'd come over from Ireland, and he didn't have any immediate work, so he set us up with him.

"He can flat out make a car body with a hammer and torch and the wheeling machine. He has a great eye for shape and form and has helped bring our operation and level of workmanship to a new level of competence.

This shop is across the street behind the main shop, and is Jimmy Kilroy's area to craft and create his phenomenal coachwork-quality metal.

"Generally, our most difficult challenge in creating these bodies is to reshape a flat piece of metal into a compound curve. And in an automobile, almost every panel contains compound curves.

"We use the English wheel and power hammer. In wheeling, we simply push and pull flat metal sheets back and forth in a machine that features pressurized rollers over interchangeable forms. It stretches the metal. In my dad's time, they still used hammers and shot bags. Wheeling allows us to stretch the center without leaving hammer marks. The old way was more time consuming, but an experienced craftsman could accomplish the same result."

Moal's love of metal "in the white"—in its raw, unadorned, and unpainted form—is evident in all stages of his creations. And he's not wedded to a sterile, smoothie look.

"When things are functional, they look good," he says. "I like to see rivets, latches, locks, handles, louvers, straps, hinges, bezels, accents. The means of attaching these components to the car are works of art in themselves. Bolts and rivets and exposed hinges add a certain texture to the whole work."

Kilroy shows some historical references.

The addition of Kilroy allowed experimentation with the venerable coachwork process of *superleggera* (Italian for "superlight"). Developed during the European *carrozzeria* epoch—a time in which hand-built motor cars went from one shop for chassis work to another, where engines and drivelines were fitted, and finally to a third where hand-formed bodies were mounted on the rolling vehicle—*superleggera* is a time-consuming and deliberate system in which a light, preformed tubular subframe matrix is affixed to the main chassis. Shaped metal body panels are then crimped to this system. The entire effort adds strength without adding additional weight to the car.

Bob Dron has been a Moal collaborator for years. The 63-year-old Harley-Davidson entrepreneur and motorcycle stylist has also been a custom car and hot rodding fool since back in the day when, defying his headmaster's orders, he'd go over the wall at his Marine military academy just to catch the bus to Oakland to polish cars so he might cadge a ticket to get inside the Roadster Show.

The Moal–Dron synergy first emerged in 1991 with creation of the landmark Heritage Royale

(continued on pg. 146)

Kilroy's metalwork is on display in Ed Adams' roadster, photographed at Bonneville.

Bob Dron's tomato-red '32 is a prime example of the Moal-Dron synergy.

(continued from pg. 143)

motorcycle, a production Softail that was gracefully sheathed in a sleek aluminum body, one articulated by industrial designer and avid street rodder Don Varner and featuring sweeping '30s-style accented rear wheel vents. The blood-red *Heritage Royale* was a show-stopper at the 1992 Oakland Grand National Roadster Show, winning the Slonaker Award for Technical Excellence and a number of best paint and design accolades. The Moal–Dron mutual *tour de force* is the *32 2 Low* tomato-red lowboy roadster.

Dron, who's never shied away from his vision—and who's never shied away from emphasizing his design points with Moal, either—says, "His old-school craftsmanship relates to the old-school building he's in. His attention to detail is very hard to find these days.

"He builds vehicles to drive, not to just look at. The '32 is basically a sprint car. Oh, it looks like an early '32 Ford hot rod for sure, but his chassis really makes that car work. This SOB really handles."

Tim Allen has commissioned two Moal cars, the first of which was his *Licorice Streak Special*. Created in 1999 and incorporating a number of emerging Moal engineering and design themes in a light, fast, rakish

(continued on pg. 150)

The wall in the background displays mementos from two of Moal's customers/collaborators: actor-comedian Tim Allen and Harley-Davidson entrepreneur Bob Dron.

The open workspaces and natural lighting at Moal Coachworks suggest a small-scale version of the workspaces of forward-thinking early-twentieth-century industrial architects like Albert Kahn.

Created in 1999, Tim Allen's rakish *Licorice Streak Special* roadster incorporated a number of emerging Moal engineering and design themes. A substantial chassis can handle plenty of power, in this case a hopped-up Ford Racing 351.

A view of the main shop area with Michael, David, and Steve Moal standing in front of Eric Zausner's new '34 Ford boattail project.

(continued from pg. 146)

black roadster, the Allen car allowed Moal and Kilroy to refine the *superleggera* process. They built a substantial, race-bred chassis that could handle 400-plus ponies roaring out of the hopped-up Ford Racing 351-inch motor equipped with SVO GT40 aluminum heads.

While engineering improvements abound in the *Licorice Streak Special*, there can be no mistaking its exterior heritage, which is in some ways a big cousin to Moal's earlier, elegant, but much smaller *California V8 Special*, now owned by another Moal aficionado, Ted Stevens. Allen saw the *V8 Special* in an article in *Street Rodder* magazine and wanted one like it.

The next car Moal created for Allen is the *Triple Nickel*, a highly modified '55 Ford two-door boulevard cruiser.

"I love his attention to and passion for detail," Allen says of Moal. "He's pretty stubborn, though he'll listen to your ideas and then present his own. We butt heads. He has his ideas about his creative stuff, and I have mine. This is just a great part of the process.

"And that shop, it's so friggin' clean. Reminds me of my shop class days—keep it neat. There's real lineage here, too. His dad's stuff was great. I was lucky enough to be there when George was alive."

Rear view of Eric Zausner's '34 boattail project.

One of Moal's most dedicated patrons is Eric Zausner, *Ferraristi*, collector (including the world's largest assemblage of gasoline-powered Spin Dizzy model racers from the '30s and '40s), historian, and astute producer of what he calls his "theme" cars, which are motoring manifestations of conjured machines from a certain period. These include the 450-horsepower, Ferrari V-12-powered *Torpedo* 12C–5.5 TS, a daring theme piece with a pronounced Italian heritage but that also integrates motorcycle and bobtail fenders and a DuVall-style split windshield. The *Torpedo* rides on a hefty, arrow-shaped, boxed chromoly steel frame with subsystems designed to fulfill a simple Moal dictum: build maximum rigidity into the framework, and let the vehicle's racecar suspension absorb road anomalies and smoothly transfer horsepower back to the blacktop.

There are other Moal creations in *Scuderia Zausner*: the *Speedway Special*, with distinct lineage to postwar circle track racers and accoutrements that might've come from an aviation war surplus store in the early '50s; a boattail '34 Zephyr; and a '36 Ford Thunderbolt. "When we were first building the

(continued on pg. 154)

19852

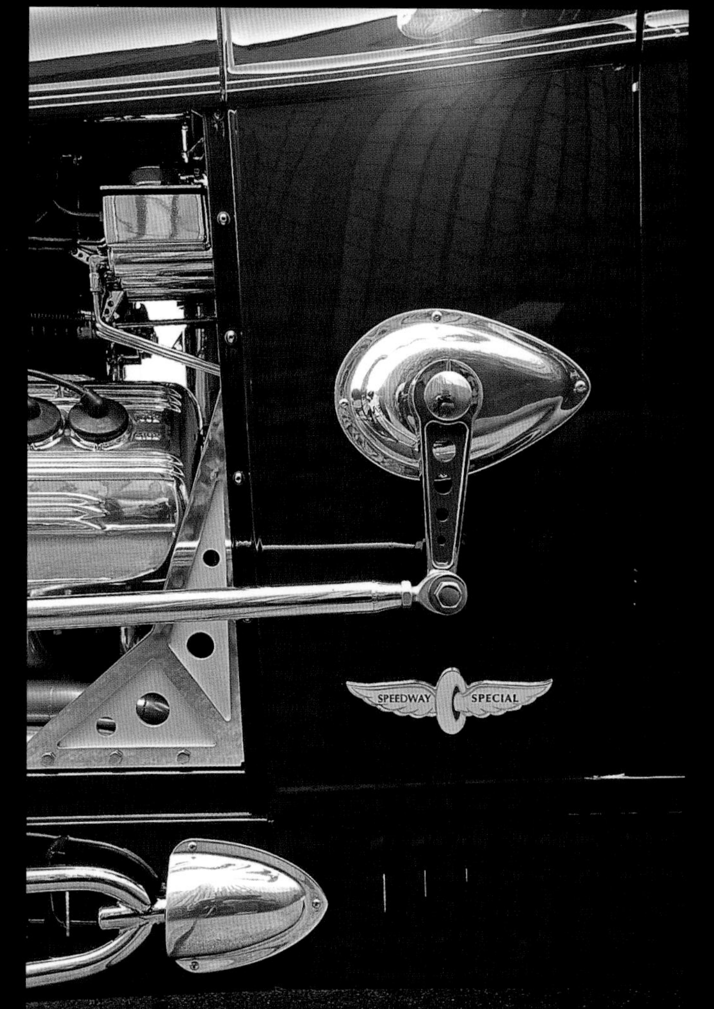
SPEEDWAY SPECIAL

MOAL

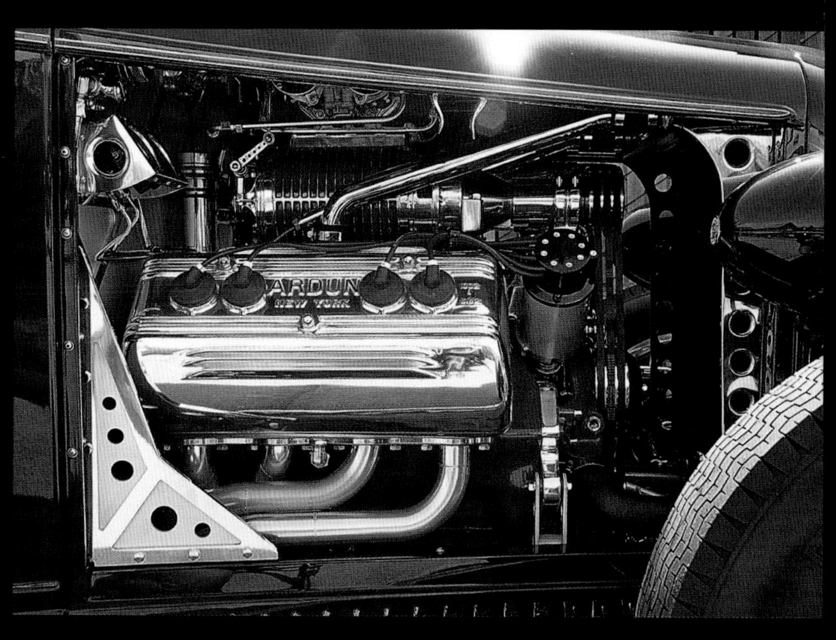
ARDUN
NEW YORK

Steve sits in a wood mockup for a new project. Mockups ensure correct ergonomics for each customer.

(continued from pg. 151)

Torpedo, Steve had me sit on a box in the cockpit," Zausner says. "He wanted to find out just how high my elbow was, how high I would sit in driving position, and he wanted to find out how wide my foot was so he could judge placement of brake, clutch, and accelerator.

"He is dedicated to the details of making it work, to work right. He addresses a unique blend of engineering, fabrication, and artistry. Some guys can do one or two of these, but he encompasses all three.

"Although he's a traditionalist, he's not stuck in a box, repeating what's been done before. He grew up with hot rods and the hot rod style of the '50s and '60s, but he also knows Bugatti, Duesenberg, Maserati—and '32 Ford. We speak the same language, and we can mutually appreciate, say, the leaf spring setup on a Bugatti.

"Our first project was the *Torpedo*. I'm sure no other shop could complete this car—it was a daunting project. Today, after driving this car, there is nothing I'd change."

Another longtime cruising buddy who's commissioned Moal to personalize his very efficient hot rod is Gary "Goodguy" Meadors, a Northern

California neighbor whose immense gatherings have set the standards for outdoor rod and custom events.

"He's a real gentleman," Meadors says. "I've known him since the late sixties, when he had a T-bucket and we'd go on rod runs. He'll take on a small job—or a complete car—and get it handled.

"For years, I had this model of a Dick Flint roadster on my coffee table at home. I just had to have a roadster along those lines, and finally (my wife) Marilyn said, 'Quit dreaming and get it done.'

"Steve was the guy to do it."

With some design collaboration with Chip Foose, Moal created Meadors' track-nosed, yellow '29 roadster.

"A drivin' son of a gun," Meadors calls it. "I've put over 50,000 miles on it—a great car. I never tire of telling anyone how beautifully this car rides and handles.

"Steve's a thinking kind of guy and a doin' kind of guy. And, yeah, his shop says what the hot rodding world's all about. There's nothing superficial here—no super-clean, lookin' pretty Chip Ganassi racin' garage. This is a workin', let's-get-shit-done garage; not something that's there to just look pretty."

(continued on pg. 161)

Gary Meadors commissioned this Dick Flint–inspired '29 roadster.

Hank Torian's 1946 "Moal Chrysler" boattail.

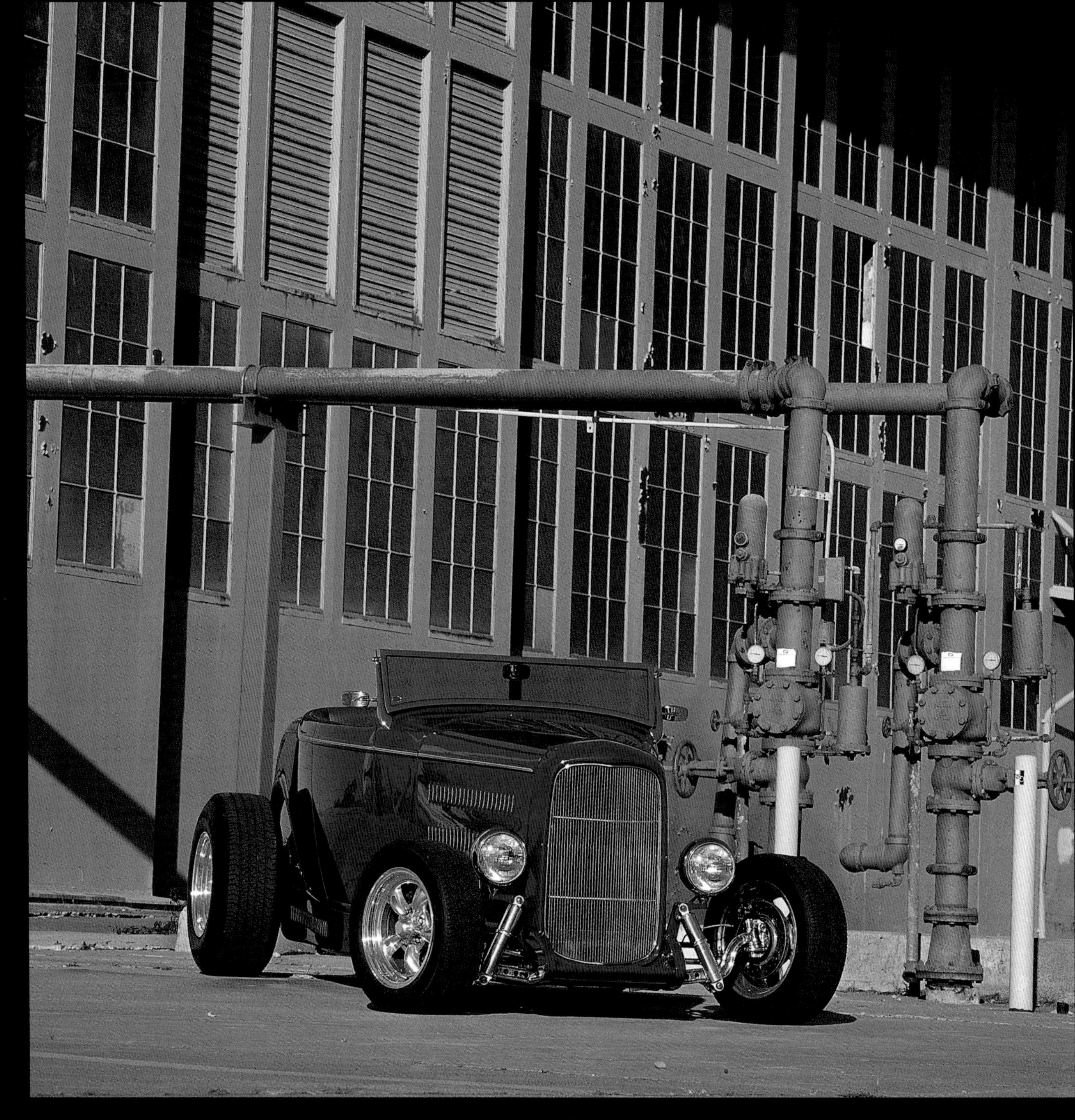

Steve's personal '32 roadster.

(continued from pg. 155)

At the Pebble Beach Concours d'Elegance—that rarified gathering of elegant and iconic marques that, for the last 57 years, has graced the greensward sweeping from the veranda of the Del Monte Lodge—collectors and coachwork *cognoscenti* are very, very serious about the proceedings. Owners are perfectionists in the ultimate, showmen and show women whose attention to detail and commitment to heritage is exacting to a finite detail.

Silicon Valley developer and noted auto collector John Mozart is one of those top-echelon challengers whose displayed vehicles attain rare perfection reflected in a precise points system. His 1938 Type 57SC Corsica Bugatti roadster won best in show at the 1998 gathering. It is a stunning work—long and low with arched pontoon fenders, swept windscreen, full body-length chromed accent moldings, exposed filler caps, wire wheels, and red leather upholstery.

"It was a one-off, built for a British colonel," says Mozart. "Steve did all the metal finishing. The car was old and was used a lot, so there were small cracks in the body. This was not a heavy restoration, but there were lots of things to fit and finish."

Having raced, built, and shown hot rods, Mozart has a broad appreciation for both European and American motoring style and form. He grew up on the San Francisco Peninsula, an area certainly given over to some serious boulevard racing and cruising—especially along El Camino Real and its pit-stop drive-ins, like Kibby's in San Mateo.

Moal has worked on two of Mozart's '32s: a roadster and a three-window. It is the hot rod work that allows Mozart to interact more creatively with the Moal team.

"On the hot rod stuff, we listen to each others' opinions," Mozart says. "In restoration, we review critical things and make decisions on the spot. Steve knows where the project is going, and I rely on him. They have a great team, and it's fun to do business with them. I enjoy my time with Michael and David and am interested in their ideas and input. Jimmy Kilroy is the best in the world. He knows what you want and understands the entire scope of the project."

The next Moal–Mozart collaboration is a Murphy-bodied 1932 Lincoln KB convertible, the last of five built. It was Edsel Ford's personal car and a debut car at the 1932 New York Auto Show. A long-lost sister to this Lincoln was driven by Edsel Ford as the pace car at the 1932 Indy race. Mozart has been trying to buy the car for 25 years and, in late 2007, purchased it at a bankruptcy auction. Moal will do all the metalwork and hardware restoration, including making a number of castings to replace broken or deteriorated components, such as the windshield posts.

"I've had them do little things, too" Mozart says, "and when I give anything to them, I know it is in good hands. Very few have these skills."

David Kelley is a professor of mechanical engineering at Stanford University and founder of IDEO, an engineering firm. When it comes to cars, he's also "undeniably an old-school guy," who began tinkering with cars when he was a 14-year-old kid in Ohio.

"My students still talk about Steve," Kelley recalls. "He came to our department a few years ago and gave a talk. He told our students, 'When you're 90 percent done with a project, that's a time that separates the great from the not-so-great. You have to find a way to re-invigorate yourself to complete the last 10 percent.

"Steve is by any measure one of the top guys in his field. He's so humble it makes it a real joy to keep coming back to this shop. He's calm under fire, and I'm sure he has a lot of demanding clients with a lot of money. His easy way is part of what makes him successful."

When asked about the old-world approach (hand, eye, and mechanical tools) in comparison to modern computer design (systems like CAD/CAM), Kelley is pretty clear.

"I think with people like Steve Moal, the real talent is in realizing one's vision; the tools can be a computer or an English wheel, a torch, or a milling machine. I think you can see who's talented, whether they're drawing on paper or using a CAD/CAM system. The most important point of view is the love of cars and the talent associated with the creation of things."

As he takes a quick look at his shop, Steve Moal waxes a bit nostalgic.

"Yes, visitors do comment about this building. It has an industrial, old-world look. It provides a good atmosphere, a creative atmosphere. It's a survivor and gets appreciated again and again. The original architects did a good job. Sort of like an old car—some designs don't fade away. This is a good example."

Eric Zausner's Ferrari-powered, handcrafted *Torpedo*. "Moal's not stuck in a box, repeating what's been done before," Zausner says. "He grew up with hot rods and the hot rod style of the '50s and '60s, but he also knows Bugatti, Duesenberg, Maserati—and '32 Ford."

"This building provides a good atmosphere, a creative atmosphere," says Moal. "The original architects did a good job. Sort of like an old car, some designs don't fade away. This is a good example."

CUSTOM
CAR
FEB 18 19 20
FAIRGROUNDS
FAST LANE
CHAMPIONSHIP
DRAG RACING
1979

11
DICK PAGE

I HAVE KNOWN DICK Page for some time and have always respected his views on hot rods and custom cars. The opening photograph of this chapter was taken in Page's shop sometime in the early '90s, and it captures the actual work involved in creating a custom car. The image shows the craftsmanship required to physically change or remodel an automobile. The craft involves many of the same elements that are used in any visual art, such as form, shape, pattern, texture, and color, all used in conjunction with the principles of design. These design principles, such as dominant feature, proportion, balance, perspective, and visual rhythm, when used judicially, spark an aesthetic response. It could be called a "gut reaction" or just plain curb appeal. In some cases, it's something deeper.

Dick and I have talked often about artistic correlations to car customizing, drawing on his long history with the culture. Page has paid his dues and has written articles for various publications. At one point, he ran a regional television show on car culture, and he has also passed along his unique expertise by teaching vocational technology courses at Foss High School in the Tacoma, Washington, area.

Here's a quote from an article called "It's Why We Do What We Do" that Dick wrote for my first book:

THE BLUE BOOK OF CUSTOM RESTYLING, *a Dan Post vintage publication, copyright 1944, revised 1952, is a great example of an early remodeling text. It encouraged the reader to invoke a 'natural sense of design and balance.' Stock production automobiles are a compromise between the stylist and production engineers, with some advertising input thrown in. Since the original contours show a lack of inspiration, a good custom can yield a distinct improvement in style and grace. Post's text goes on to say that every car is subject to optical allusions, some good, some bad. Customizing, if it is to be successful, according to Post, should highlight the finer values and camouflage the less attractive ones. The book is hard to find but recommended reading for anyone interested in why we do what we do. Most modifications have the goal of making the car appear longer, lower, and wider. Early customs were often the owners' sole means of transportation, and today's on-the-ground stance was not an option. In my never humble opinion, a custom car dropped too far detracts from the grace of design by disrupting the smooth total-package look. It becomes the focal point and takes away from the design.*

Dick Page working on Denny Hall's '51 custom Merc, Tacoma, Washington.

Denny Hall's '51 custom Merc, which he has since sold to Brenda and Dave Kreg.

The visual point of the racer's "form follows function" mindset really influenced the custom crowd. The chopped-top coupes and sedans (with stock engines) looked fast just sitting still. Lots of '33 to '36 customs were built. When Detroit iron got big in the 1940s and 1950s, the Caddys, Mercs, and the like were too heavy to be race-car fodder, but the old styling practices stayed popular. Racer tricks, like shaving lumpy door handles and hood and trunk trim, and the ultimate wind-cheating practice of chopping tops stuck with us.

The second photograph is of the same '51 custom Merc as that shown in the first photograph, at the time owned by Denny Hall, in a further stage of construction. We set up the evening at Dick's shop, with the Merc inside, and Dick's well-known chopped and blown '32 Tudor sedan (with way over 120,000 street miles on it) and Hall's '29 roadster pickup out in front. Denny has long since sold the Merc.

Dick still has the chopped '32 Tudor. In 1990, he claimed that he had less than $3,000 invested in it. Dick was into the retro, or nostalgic, movement before it was popular. Dick is one of the originals. It is not a question of fads or movements to him. He has a true sense of hot rodding's roots, and a long history of involvement in the hot rod and custom car culture.

Denny Hall's '51 Merc project and Dick's well-known '32 chopped and blown Ford Tudor alongside Denny's '29 Ford roadster pickup. Denny and Dick are placing the rear fender skirt on the '51.

Dick Page was into the retro movement before it was popular. In 1990, he told me he had less than $3,000 invested in his '32 Tudor. At the time, he had already racked up more than 120,000 miles in the car since buying it in 1974.

The '77 four-bolt Chevy truck block in Dick's Tudor has a 4-71 Jimmy blower topped off with dual four-barrel Carter carbs (which replaced an earlier Predator 6000 carb) and a Hilborn scoop.

12
THE ROLLING BONES

I RECEIVED A PHONE CALL from Ken Schmidt sometime in 2001. Ken wanted to get together to do a photographic session out in the western desert. The Rolling Bones (Ken Schmidt and Keith Cornell) were planning a trip out from their home state of New York, bringing with them the two '32 coupes that soon became known as the "Wicked Sisters."

My first reaction was to ask Ken to send out some photographs of the cars he was describing. Something in the way he talked made me think that they might actually be serious about driving vintage-styled hot rods across this beautiful country. He was describing Ford flathead- and Y-block-powered hot rods with early Ford brakes, Schroeder sprint car steering, Halibrand quick-change rearends, no cushy interiors, lots of louvers, and—an unusual choice for a cross-country trip—skinny steel wheels with bias-ply Firestones. They also were running generators on the cars, not alternators or modern electrical setups. Oh yeah, and no mufflers, but rather straight pipes.

I mentioned the cars to Mark Morton at *Hop Up*, as we had been talking about what to set up for feature shoots for the next *Hop Up* annual. He mentioned getting some other cars together for the session, all vintage '32 Ford hot rods. No flashy paint, no billet, and no modern drivetrains or suspensions.

Mark had five other '32s set up to join us, all meeting at Bill Carey's used car lot in San Bernardino. It turned out to be one of the best photographic sessions that I've had, and the day was perfect for all concerned (except for a blown tranny in Billy Vinther's '32 full-fendered coupe). It was a joining of the East and West Coasts, as Ken and Keith's cars fit in and without question added an interesting element to the session. Better yet, everybody really got along and all have since become good friends.

Ken's red oxide–primered coupe has essentially remained the same since then. Keith, on the other hand, changed his coupe's body, as he wasn't completely happy with the top chop and a few other things. The main difference was that Ken's coupe had been chopped a half-inch lower in the front, creating a slight wedge to the overall look.

Keith sold his body and found a very nice, complete, '32 three-window in a barn back east. He whacked another inch out of the top, pushing it just a little lower than Ken's coupe, but with the same half-inch difference between the front and back.

Riding in a heavily chopped car for long periods of time—and I mean really long—visibility is a factor.

The Bones work as well from the inside as they do looking at the visuals from the outside in. The proportions worked visually at every angle. A lot of that is due to the very astute care that both Ken and Keith take. They bring the front posts back to the top, which on a chop that severe seems to make a difference. Ken's background in art and his understanding of aesthetics comes into play in all this.

Ken Schmidt and Keith Cornell, Greenfield Center, New York.

Ken Schmidt's '32 Ford roadster, "like it might have been in '49."

The cars of the Rolling Bones boys suit my beliefs of what a hot rod is.

The Rolling Bones crew showed up out on the salt in 2003 and they guys were hooked by the salt fever that brings us back year after year. They decided at that point that the '32 roadster that Ken was building would become a racecar.

In 2004, my wife, Kim, my son, Nathan, and I met the Rolling Bones in Denver after they had crossed the country from New York. At that point, we took Ken's coupe for just over a week to drive through Moab and Monument Valley and on to Los Angeles for the L.A. Roadster Show. Photographing the three cars (including Ken's new '32 roadster) in Monument Valley was a treat.

In 2005, I flew to New York to drive out to Bonneville with the Rolling Bones crew, which turned into a very memorable trip. The '32 roadster was driven across country with many different drivers, including me, and it raced with the same drivetrain after being converted to racecar trim in the parking lot of the Nugget Casino in Wendover, Nevada. The scene recalled earlier

Ken Schmidt's '32 Ford coupe gets a push at Bonneville.

Bonneville times, when rebuilds and repairs took place in the parking lots of area motels.

The Bonneville adventure in 2005 ended with a drive off the salt in 6 inches of water after a late-afternoon Tuesday storm that shut down the races for the rest of the week. Ken and Keith were back again in 2006 to race the roadster, but they lost the engine early in the week. In 2007, we met at Bonneville, and Keith and Ken gave my wife, Kim, the opportunity to pilot the roadster down the salt on an official pass at 115 miles per hour. She's now fully committed and fired up to drive again. Having driven the salt myself, I can see why it can become one of those lifetime obsessions.

I have the utmost respect for the values represented by Ken and Keith (and the rest of the Rolling Bones crew). They are doing this for the fun and pure passion of the experience. To climb into real, vintage-style hot rods and make that many cross-country trips says two things: they are passionate about what they do, but they also can build the cars to make the trip time after time.

VIN'S
32
VJS·10
193
C

Five Deuces inside Keith Cornell's Rolling Bones shop. From left: (barely visible) Ken's '32 coupe, Keith's '32 coupe, Dennis Varni's '32 Tudor, Ken's '32 roadster, and, barely visible behind Varni's Tudor, another Tudor project belonging to Mike Manno.

Experiences and Influences . . . from the Beginning

By Ken Schmidt, buildup photos by Rolling Bones

All hot rodders have had experiences in their past that shaped and influenced them. It might have been when some young kid in a roadster stopped and asked if you needed a ride. Maybe it was that tough kid with the rolled-up sleeves and the D-A haircut driving the primered coupe with the loud mufflers. Or, possibly, it was the Tudor you saw in the little pages magazine tucked inside your textbook during history class. You knew hot rods were dangerous. Your parents hated them, and you couldn't wait to build one of your own.

There was nothing like actually building your own hot rod, with all the experiences and learning included. You didn't care how rough, crude, or even plain ugly it was; in your eyes, it was just so cool! You couldn't wait to show it off. We have all been there, improving and learning with each one.

Then comes the day, the day you see the "one." For Keith and me it was the Doane Spencer '32 roadster. It was the quintessential hot rod, and it was way ahead of its time and way ahead of the curve. When you look at the hot rods we build, it becomes obvious that we have stolen most of our ideas from it.

I had an older brother who was born in 1940, and since both our parents had to work, he was stuck with me tagging along. At 16 he looked 18, and he was the one your mother warned you about. It was 1956, and kids were car crazy. He was driving a hot rod, breaking every law he could and keeping our parents awake every night. I was 10, with my brother for a hero, and I was going along for the ride. We hot rodded together for the next 40 years.

Keith grew up around cars. His father loved the cars of the mid-'50s. After graduating from high school, his older brother started a repair shop in the barn across from their house. Keith went through a string of muscle cars until he traded into a glass Deuce roadster powered by an old and tired flathead. It was like getting that first kiss from the beautiful girl next door—he liked it!

About 15 years ago, my 16-year-old son and I took his sort-of-finished, 327-powered, 1949 Ford truck to a weekend car show. It was there that I first crossed paths with Keith. He was driving his nasty-sounding, flathead-powered, black '34 Ford roadster. It was the coolest, and to me, it was the only real hot rod there.

It wasn't until four or five years later that we started running together. He was building a Deuce three-window for a friend, and Keith told him if he ever lost interest in the car that he wanted first dibs. His friend changed directions, and Keith ended up with the coupe. I was looking for a Deuce three-window myself, and with Keith's help, we found one.

We never planned on building them as a pair of "Wicked Sisters," or, for that matter, building hot rods for others. However, in working together, we soon realized that we shared many of the same ideas and a deep appreciation for postwar hot rods that served as daily transportation, only to terrorize the dry lakes on the weekends.

There is no rule book for building hot rods, and we built those coupes for ourselves. While some around us snickered, to us those cars were the pure essence of what a hot rod is, or was for that matter. Neither Keith nor I are parking-lot, sit-in-a-chair-with-a-duster-in-hand, hoping-for-a-plastic-trophy-at-3 p.m. street rodders. But still, after driving them across country for the L.A. Roadster Show, we never expected the response they got.

Turning down one purchase offer after another, people started asking, "Would you build me one like it?"

Keith and I worked well together and respected each others' strengths. We looked at each other and said, "Why not?" Best-laid plans and 3,000 miles can make interest in having a hot rod built by a couple of guys working out of an old barn in upstate New York wane and die on the vine. No matter; we were on to new adventures. Bonneville was calling, and a trip was planned, knowing that the '32 roadster in progress in that old barn could easily change directions and become a combination street and race hot rod.

Two years later, we were back in L.A. for the Father's Day event. We were parked in the swap meet, and this guy with an easy smile came up and started telling us how much he liked our coupes. A while later, he came back and said, "I've had this Peter Vincent calendar in front of my desk for over a year. I love that cover picture of your coupe on the El Mirage dry lake so much that I have never even opened it."

He came back a third time, and that time he introduced himself as Dennis Varni. Next thing we know, he was buying parts and making

A top three-quarter bare-metal view of the Rolling Bones '34 Ford three-window created for George Poteet.

arrangements to fly back east to look at Keith's Tudor body. He had this '54 baby Hemi for it, and he dropped off a quick-change rearend for us to take home. Our heads were spinning.

Three weeks later, Dennis was in our shop looking over the body and frame. Within an hour, we were shaking hands and he said, "Let's build this." The hot rod god "Deuce" was watching over us. None of the other people who had approached us would have been the right customer for us to start with.

The reasons for and the approach to building hot rods can vary greatly. Certainly those who build cars for others started out with a love for them, but running a business in order to make a living can often compromise your goals. Jobs that may not stir your creative juices have to be taken. Payrolls have to be met, and bills have to be paid. Keith and I are in a unique situation. Right from the start, guys wanted us to build them in our style of hot rod. Our goal is not, and never will be, to have a street rod factory. In fact, we want nothing to do with that rat race. We only build hot rods in our style, period!

Now, defining the term "hot rod" is about as elusive as finding that original Deuce roadster that has been stored in some old farmer's barn since April 1, 1932. There is no doubt that the Deuce roadster is the foundation. In fact, in the beginning, coupes and sedans were not even considered hot rods, much less allowed to compete on the lakes. But, there is something ominous, if not evil, about both coupes and sedans. Of course, we are not talking about the street rods built for cruising on Sunday afternoon. We are talking about the nasty ones. The ones that seem to come out of the dark like a nightmare, with their tops chopped severely and their windshields laid back. Once in the open, they seem like panthers with their low-slung headlights emitting a sinister glow.

Keith and I build hot rods in that same old barn his brother started out in all those years ago. Starting out on a dirt floor, Duane (Keith's brother) poured sections of concrete as his business grew. Discarded windows, a discarded overhead door, insulation, upgraded wiring, and even some sheetrock were added over the years. In fact, the milking stanchions are still in the back alley with the hayloft overhead. We do have running water, and someday we may even have a bathroom. It's a place no self-respecting street rod shop would even consider. To us, though, it is perfect. With limited tools, and as some would say, limited skills, we walk through the door each day and step back into a simpler time.

For us, it is about using many of the same parts and stealing the best ideas of those we respect, and while these early hot rodders had no interest in the history they were creating, we do. They were simply trying to go fast with what was available. We are trying to capture the look and the soul of the late-'40s/early-'50s hot rod.

There seems to be a huge debate over this patina craze, and, like any trend, it came about because someone, or some magazine, said it's the cool thing to do. A good friend and artist who made the best American Indian "artifakes" once told me, "Anyone can make something dirty, but it takes an artist to make something look old."

Notice the shortened and simplified tail on the '34, and, of course, more louvers.

It is the same with re-creating a vintage hot rod. The right fake patina (that has so many rodders' shorts twisted up in a bunch) will help instill the look and the soul. No one in our shop cares if someone else likes the look or not. When we sit down to have lunch, we don't look at our project and imagine a street rod sitting in line with other street rods, all shined up with no place to go. We look at the hot rod we are building and imagine the driver hunched down with his hand gripping the wheel, engine screaming, and the dust plume trailing as it raced across the dry lakes some 60 years ago.

If you don't get it, who cares, because we do! We all remember that first day we started building our first hot rod. Enthusiasm was at a high point as you and a couple of friends started ripping off pieces and parts, stripping the car down to its bare essentials in an effort to make it look cool and be fast. Keith and I have never really strayed far from that path. We try to build hot rods with that same simplicity today, boiling them down to their purest forms, where they have everything they need and nothing they do not.

Great hot rods are works of art, and we will argue that point until the beer runs out. Just like an impressionistic painting, when you look at a great hot rod it will tell you just enough of its story to grab your emotions and then let your imagination take you as far and fast as you want to go.

That is what we try to build—stories in the Rolling Bones style—one at a time.

It is sometimes a little intimidating when we hear people talk about the "Rolling Bones style." We would love to take the credit for it, but the truth is that it belongs to those from the past. Keith and I follow one basic principle: nothing should slow it down visually. And we try not to screw it up.

To understand what we mean, let's start with the premise that hot rods are basically cars that someone has modified in an attempt to go faster. That may seem awfully simplistic, but let's also assume for the sake of argument that you can break down the building of a hot rod/racecar of the past into two simple parts. First, the builder took off every unnecessary part. Less weight equals more speed, plain and simple. The problem then became how to beat the other guy who did the same thing. More horsepower, that's how!

And that's about it, except for that one often-misunderstood principle of form follows function. It's about line and proportion; in the quest for speed, those who can figure out how to cheat the wind more often than not come out on top. I'm sure there were a lot of late nights in the garage during the week after the first hot rodder showed up at the meet with a chopped top or a track nose.

Nostalgia has a way of making things seem a little better than maybe they really were. Not all the hot rods of the past were half as cool as we remember. Just like those who build cars today, there were only certain few who had that magic touch and who could not only make their hot rods go fast, but could make them look fast. Funny thing about that: the ones that looked fast usually were!

Traces of silver metalflake and several other colors in the door jambs and a 3-inch chop put a smile on your face when you think about her history. Once a racecar, as evidenced by the remains of a roll bar, you have to wonder if she once raised the dust of El Mirage, tasted the salt of Bonneville, or smoked her tires when the flag dropped. What stories she—and those who owned and raced her—must have to tell.

The Start of the Build

While there can be no doubt the Doane Spencer roadster has played an important part in what seems to have become the Rolling Bones style, two other historic hot rods also have contributed to our look. Has there ever been a nastier coupe than the Pierson brothers' '34 three-window? The first to lean the posts back, the Pierson brothers set the bar at a level that few others, with the exception of the *So-Cal Coupe*, have come close to. Name any list of nasty, mean, wicked, or just plain badass competition/street coupes, and if those two coupes are not lurking at the head of your list, it's incomplete.

Keith and I have been fortunate to have built hot rods that seem to have struck a chord with hot rodders who also have the same appreciation for the history of this truly unique American sport/hobby of ours. One such individual is George Poteet, who is an avid hot rodder but is also a serious land speed racer who holds several world records at Bonneville. Already in the process of building him a Deuce five-window, we mentioned in conversation that we would really like to build our version of an almost street-legal version of the Pierson brothers and So-Cal coupes. His next words were, "I've got a body. When do you want it?"

Once again "Deuce" had smiled down upon us. True to his word, George's '34 three-window body was sitting on the floor of our old barn in a matter of weeks.

While most of the bodies that show up at our door are un-hot rodded, they still have 75 years of dents, rust, rips, tears, and general abuse on them. Keith's own coupe had a short life as a stock car at county fairs back in the mid-'50s, until someone hit it in the back, broke the spreader bar, and punched a hole in the gas tank. It sat in a barn after that. In fact, it still had its brushed-on yellow paint job with red racing numbers when it finally saw the light of day. My coupe had been made into a truck by throwing away the deck lid and cutting the back open, which was a common practice among farmers years ago. George's three-window was different; his came with an attitude!

The Chop

The chop on a car can make a bold statement. For us, the more attitude the better, but certainly you can chop a car too much. We have all seen cars that have become cartoons. The key to making a serious chop look right is as much about how it relates to the proportions of the rest of the car as it is about how deep the cut is. By laying back the windshield and giving it a slight wedge, you accomplish several things. Again, it's the form and function deal. The function is to allow the wind an easier path over the top. Form comes into play when making adjustments so that the wedge is correct with the rest of the lines, making the car's profile race forward. It also keeps the length more in proportion with the body because a filler piece is not needed to reach the post.

Once the body is squared up and the doors are fit, bracing is added to hold everything in position. The lead from the original chop was removed before we laid out an additional 4 inches of chop.

This is not for the faint of heart!

The lay-back on the first chop started about 2 inches above the cowl, which means when the windshield is closed, the bottom corners will stick out. Rather than try to redo the bottoms of the posts, we cut off the cowl eyebrow and lower post section and started with another. That way the windshield will close as Henry intended.

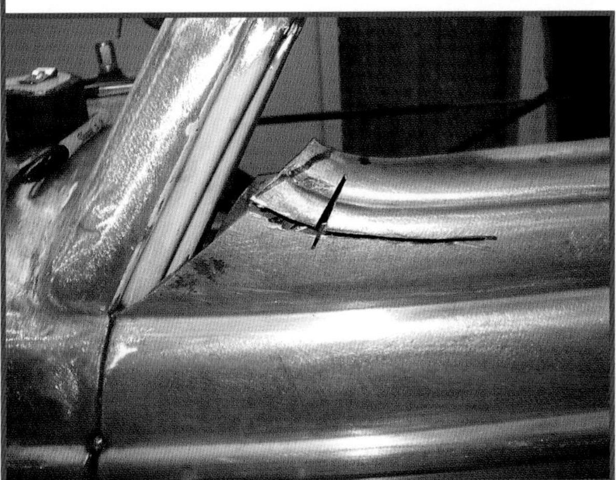

With the new cowl section in place, the posts will form a straight line. Next, the door is cut from the top of the reveal to the curve in the lower window opening. With the posts laid back about 30 degrees, two relief cuts are necessary.

Time for the doctor to start sewing her back up.

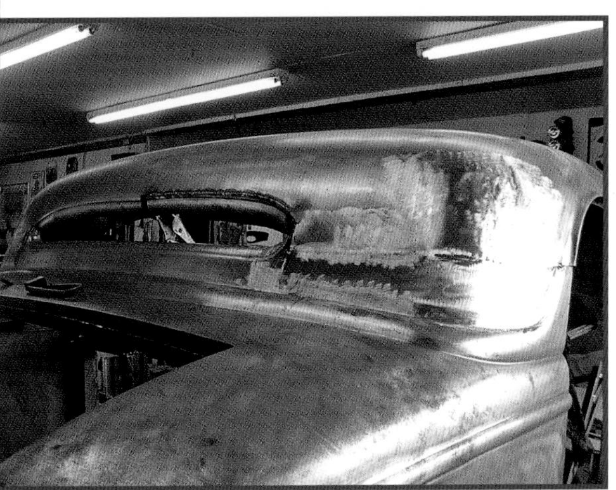

This looks a little better.

The Rear Pan and Wheel Placement

Have you ever noticed how a stock gas tank hangs down like a weight, and the rear frame horns fight the forward-looking lines you are after? Not to mention that they add length to the wrong end of the car. Talk about slowing the car down visually! Get rid of the horns, or shorten them and bring them in tight. Put the gas tank in the trunk, where it belongs. We have all heard the phrase, "It looks like it's going a hundred sitting still." Just making this change will get you up to about 50. It shortens the distance from the back of the car to the center of the rear wheel. This, along with not lengthening the top proportionally, lengthens the distance to the front, setting it in motion.

You must also center the rear wheel and tire in the wheelwell. Sometimes it's the simplest change that does it, but they all must work together to achieve the look.

On a '34, there is the problem of the ducktail shape behind the rear wheel. Some may think it looks cool, but again, we feel it lengthens the wrong end. Think about a sprinter: are his legs stretched out behind him or are they up underneath so he can explode out of the blocks?

A pie cut is made above the mounting holes for the rear panel, and a relief cut is made along the edge of the mounting plate. Basically, the same cuts were also made along the outside of the body. Next, new rear corners were formed.

It took surprisingly little work to tuck the corners in and match them up with the panel below the deck lid. It may seem like a contradiction to streamline an early Ford hot rod, but that's exactly what we are trying to accomplish. Left alone, the ducktail will let your eye run off the back, stopping the look of the car's forward motion. While the rolled rear pan keeps your eye in motion, it also provides the perfect solution to what becomes the downfall of many hot rods—where to put the taillights, exhaust pipes, and license plate. Keeping in mind that nothing should look like an afterthought, we mounted the taillights in the pan and ran the exhaust along with the push bar, which will double as the plate holder with the all-important Rolling Bones topper out the back of it.

With the buck as our guide, we formed the rolled rear pan.

Once everything was laid out, the passage holes for the exhaust and the push bar were cut in. Next, a cover for the push bar was made from a piece of exhaust pipe and was added to the bottom of the pan. Looking down into the pan, you can see the exhaust passing through. The push bar is mounted to the bracket that will push against the rear crossmember.

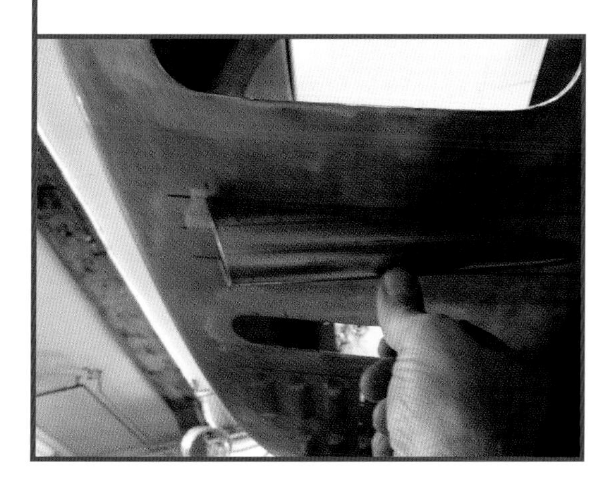

Of course, we all took our first hot rods up and down the road in front of our parents' house for a little test run before they were legal to drive on the road. Now, I don't know about you, but when the big day finally came, the only thing on my mind was taking it screaming down the road to show my buddies. I never once thought about where to put that little thing called a license plate.

This is what a lot of guys will see when George screams by them. When we got it, the body had been poorly chopped about 3 inches, which of course was not near enough. We just kept on cutting until she looked right. We are not sure exactly where she ended up, but we would guess at least 7 inches of chop. We did widen the back window a little just to make her a little nastier.

It's a mystery to us why so many guys overlook something so basic to the look as the correct placement and size of the rear wheels and tires. We believe the tire should be centered in the wheelwell, and there should be about 2 inches of space around the tire and the body reveal.

Once the rolled rear pan is attached, when your eye runs toward the back, along the top of the body, it will round the corner and race forward, toward the front. It loses both speed and direction at the bottom of the cowl while traveling up the curved line between the inner fender panel and the bottom of the hood.

Remembering the principle that nothing should slow it down visually and, of course, trying not to screw it up, we decided to lay the bottom of the cowl down. First, we cut the reveal off the bottom of the cowl. Then, using two '32 five-window patch panels, we laid the cowl down.

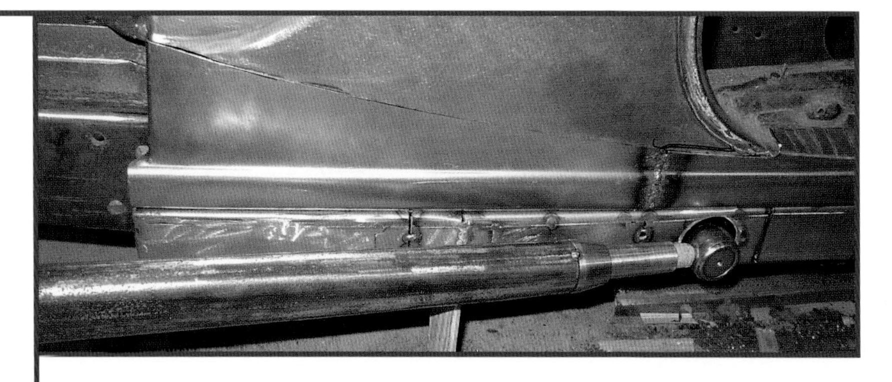

Bang zoom. With the cowl taken care of, we directed our attention to the hood. Soon we realized that we had to figure out the under-the-hood panel first. When we build a hot rod, we try to use as many old parts as possible but certainly not always in the manner they were intended. The back section of our old barn/shop is littered with old parts. In the pile we found a couple of original '32 hood tops with the back inch or two cut off. They looked like under-the-hood panels on a '34 three-window highboy to us.

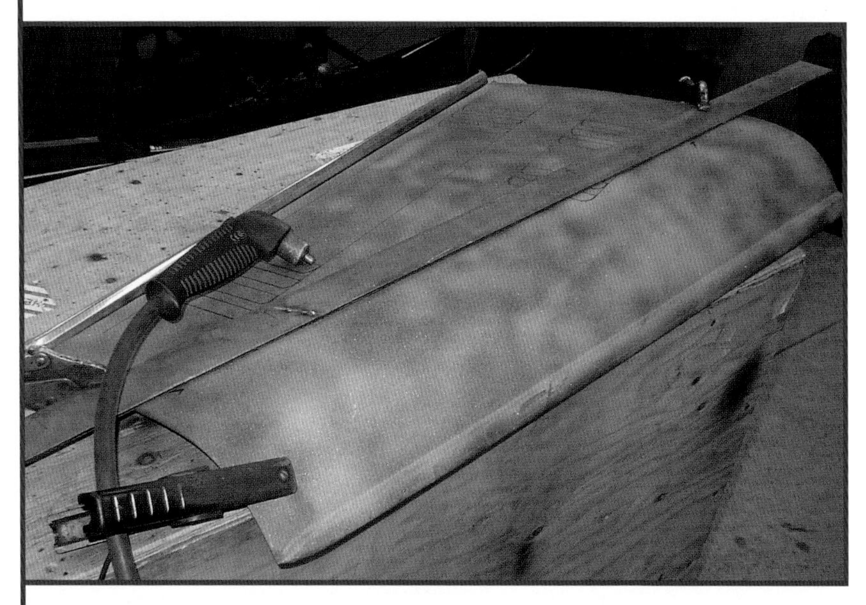

It just takes a little imagination.

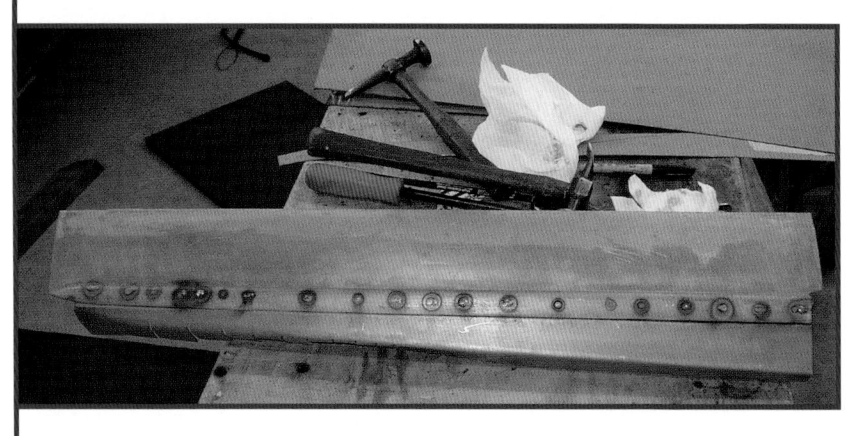

Using the lower side, with the spear for the top and the hinge side for the bottom, we came up with this. Wanting the lines of the car to pick up speed, and starting at the front of the panel with the original hood spear, we widened the reveal slightly as it flows all the way back to the rear pan under the tail. It was necessary to split the cowl panel to make both panels flow together.

Mocked together you can see the spear getting wider as it flows into the cowl. The lower panel also will widen as it flows into the belly pan/frame covers. Lines like these make your eye pick up speed as you look at the car.

No Bondo Buckets!

Bondo bucket: a derogatory term from the past if there ever was one. It's an easy product to use, though, and the truth is that the plastic fillers made today are pretty good. Lead costs more money, takes more work, is certainly more difficult to master, and who will know the difference when she is done? We will, and she will; it's part of her soul!

After thoroughly cleaning and tinning the surface, the lead is applied using a mild flame.

An untinned metal shield is placed between the door and hinge in order to keep a nice gap.

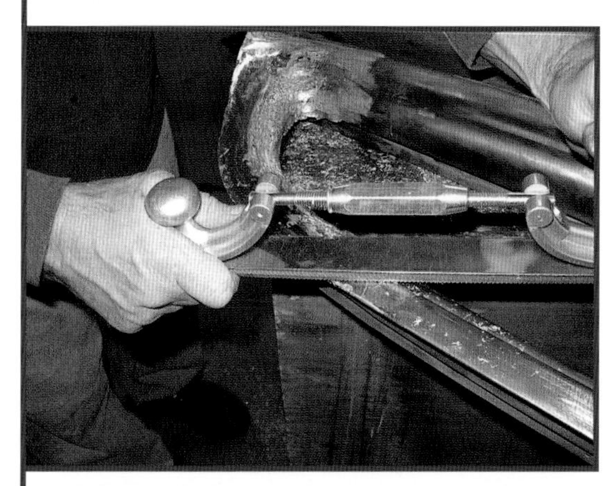

There is quite a bit of work in reshaping the doors after leaning back the posts and chopping them 7-plus inches. After the lead is applied, it's time to get out the files, coarse sandpaper, and patterns to make sure both doors match.

With the frame cover/belly pan attached, and using several layers of tinfoil for the gap, the lead is roughed in so both surfaces match.

After matching up the surfaces of the body reveal and the frame cover/belly pan, a little finish work is completed.

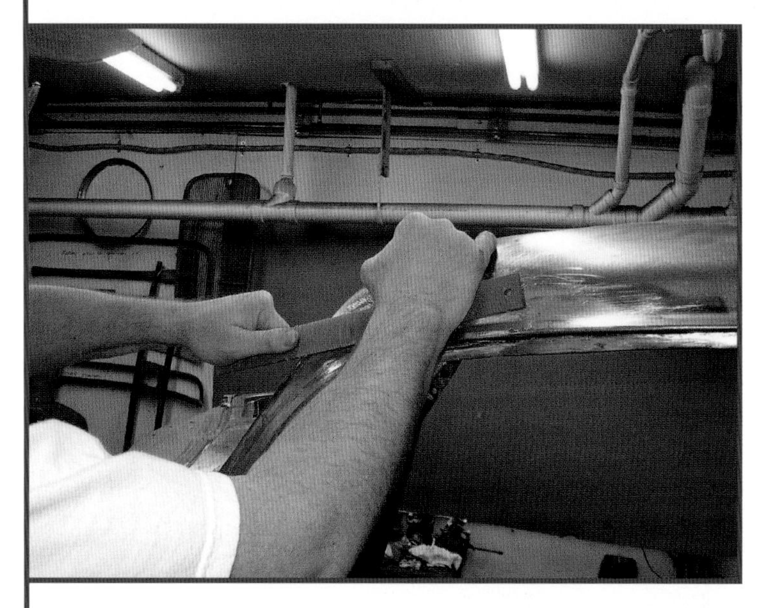

There is no drip rail on '34s, but there is a hooded reveal line over the doors that needed a little lead work.

All the work is worth every bit of it, and this says "hot rod" to us.

The Top Insert

It started with my Deuce three-window. I wanted to save the original canvas insert like Keith's coupe, but it was just too far gone. Reverting back to the foolishness of youth, I ripped it out and threw it away, including the "chicken-wire" base. We decided rather than the typical welded-in top, a removable steel insert would look cool.

We have to laugh when we hear all the reasons people think of for the louvers we do—everything from, "They put a tray with drain tubes underneath," or, "They're just there for looks, 'cause they have a solid panel under them."

These cars are meant to be driven; they are loud, nasty, raw, and pure. The louvers are the low setting on the air conditioning. Open the cowl vent for medium, and crank out the windshield for high. The configuration of the tack strip and the channel in which it fits on a '33–'34 coupe makes putting in a removable insert much easier than on the '32. The first step was to flatten out the base of this channel.

Next, the inner lip had to be ground down, allowing a flush fit for the outside edge. After bolting the tack strip in place, we welded the heads to the strip.

When we built the insert for Dennis Varni's sedan, we used the roof of an early-'50s Jeep wagon, which proved to be fairly labor intensive and expensive. Now we order the panels from Walden's Speed Shop. They have the perfect crown, no rust, no dents, and need no shaping. It's a no-brainer. The insert comes several inches bigger than needed. Clamping the tack strip on it creates a perfect pattern for the plasma cutter to follow.

At this point, the insert is slightly too big for the hole, so care must be taken when we bolt in the tack strip, making sure it sits exactly flush with the top of the body by using washers under each bolt.

Looking like some sort of torture device, the insert is glued to the channel. With the washers removed, and once the glue has set, the insert will be hand-filed for the final fit.

The Frame

There is something that appeals to us about using the original frame when building these hot rods. Truth be known, the re-pops are pretty good, less work, and a hell of a lot cheaper all the way around, but they are not the original "bones." Starting with a set of fairly well-abused originals and not much of a plan, we figured we couldn't hurt them much more than they already had been.

Stock rails start out with a pretty good upward curve that comes back down when the rails meet the firewall. Because we were going to eliminate all that inner fender bubble and curved hood-bottom confusion, lengthen the wheelbase about 4 inches, and replace the stock grille with a track-style nose, our first move was to flatten the rails and pinch them. We then tacked in Model A front and rear crossmembers.

We figure if your hot rod doesn't bottom out once in a while, it ain't low enough. But after sticking a front end and rearend underneath, it was obvious that dragging the oil pan on the road might not be a good idea.

We ended up using a '32 front crossmember because we needed to pick the front up. We changed the arch of the Model A rear crossmember because we needed to drop the back. It's all simple stuff, but it's a must if you want your hot rod to sit right.

Once sure of our crossmember placement, it was time to plan the mounts for both the front and rear split wishbones. To us, there is nothing worse than a bunch of brackets sticking up, out, or hanging down. It shows the builder took the easy way out, planned poorly, or both.

A '34 frame has a nice center X-member already in place. Using the same basic idea that we use on our Deuce frames, we inlet our mounts into the boxing plate and through the center "X" for the rear bars. The round tunnel in the boxing plate is the backside access hole to the tie rod end for the front wishbone.

The front wishbones will be mounted through the sides of the frame covers into a tapered bung. Once in place, there is no way to remove them without wrecking the frame covers, so we made this press. The same bolts will also hold the protective cover.

Detail shots of the rear of the frame; the bolts in the "C" will mount the rearend housing bumpers, which will soften the blow when she bottoms out.

Stealing ideas we have seen along the way (if anyone tells you they haven't, they are lying, 'cause it's all been done before), we use '39–'40 rear wishbones, a length of threaded rod, a gusset plate, and a couple of rod ends to make our rear bars. Mounted to the front side of the housing at an angle and tucked up in the frame, they work great. Even better, unless you get down on your knees and look up, you won't even see them.

More than one street rodder has asked us how we get away without running rear shocks. Of course, we do run rear shocks; they're mounted to the inside of the boxing plate with the dog bone receiver welded to the top of the housing. They are impossible to see without crawling under the car, and we build them low enough so that's pretty hard in itself.

Nothing should slow the car down visually.

Hood, Nose, and Door Hinges

Once the panels under the hood were fabricated, it was time to move on to the hood and nose. First, we dug out an old '34 hood from the pile out back to use as a starting point. Then out came the cardboard and wood. Before long, it started to look like something. After narrowing down the shape, we started making the buck for the nose.

With the buck finished and the grille opening determined, I looked at Keith and said, "You make it." Then he looked at me and said, "No, you make it." Then we both said, "Let's get Terry to make it," as in master metalman Terry Hegman.

Knowing the quality of Terry's work, we were confident enough to louver a stock '34 hood top with no radiator cutout. All we had to do was trim and file-fit the front edge. Starting with hood side blanks, we trimmed and bent them to fit (the payoff for the hours of planning, measuring, and taping pieces of cardboard together to mock it up).

The chop, reshaping the tail, adding the lower panel in the back, the louvered frame covers, the panel under the hood, the nose, the hood, the louvered roof panel, and the longer wheelbase all work together to get the look we are after. In the end, however, the simple things are what separate one hot rod from the pack. On a three-window, the simplest of all for us may be the removal of the center door hinge. Sure, it's easier to take off the top hinge, but the three-window doors are heavy, and sooner or later the door will start to sag.

Front Friction Shocks

Barely street-legal, everything about this hot rod screams "land speed racecar." Built to cheat the wind, the frame is covered completely. Deciding what to use for front shocks and figuring out how to mount them was a problem. Occasionally, we will pop the tops off of a couple beverages while standing around looking at her at the end of the day. Sometimes those caps are pretty tight and it takes a lot of "friction" to twist them off. Friction . . . that's it! Let's put friction shocks on her! We figure anytime you can use old parts to make parts that you want to look old, you're ahead of the game. Starting with a round disk with a hole in the center, an old shock arm, cut-down dog bones, and a bent-up piece of flat stock, we were on our way. With the old arm welded to the new bent-up flat stock, we started to give it a little style.

Using the front crossmember as the base, we welded in a disc at each end. Next, two brake/clutch material discs were sandwiched between the shock arm disc and the outside disc. Then the front crossmember was welded in.

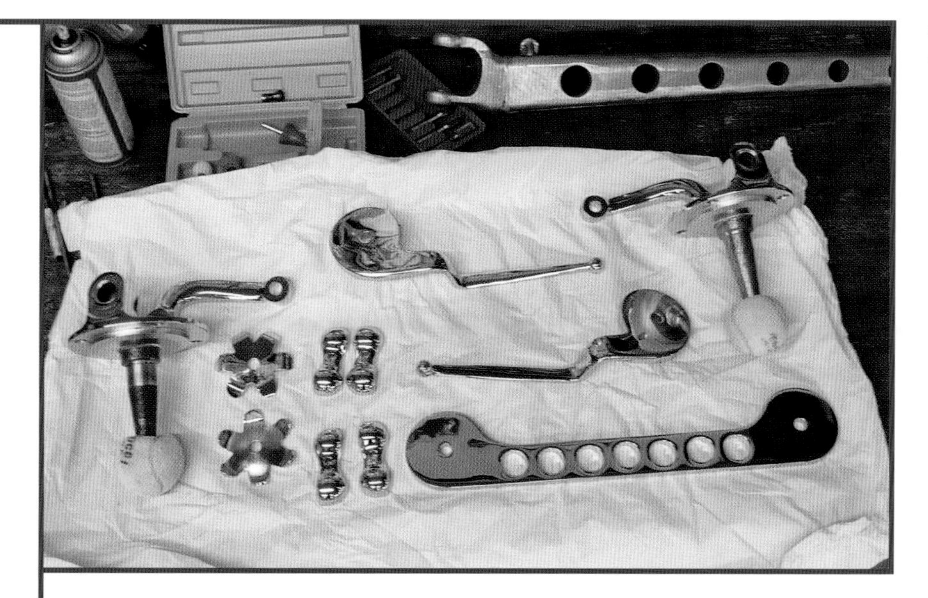

Ain't they pretty? Don't worry; we'll age them down when we're done.

Mounting the first set of discs to the crossmember and then having the double disc plate across the front allows us to have two friction discs in each shock. When she is all together, if you look through the bottom of the grille opening, you will see the chrome front plate. All you will see on the outside are the old shock arms and the cut-down dog bones.

Mounting the Headlights

In the beginning, most guys drove their hot rods to the lakes to race. The headlights were one of the first things stripped off, as they tried to make their cars more aerodynamic. Sooner or later, the hot rods we build seem to end up on the salt. With the nasty chop, the sound of a vintage engine with a straight exhaust and racing numbers on the doors, removing the headlights just adds to the flavor.

Keith and I will readily admit to taking the headlights off our coupes during Speed Week. We use our coupes as chase cars when we race our roadster. More than once, an official has come running after us yelling, " Stop! You can only drive your racecar on the track!" And that, boys and girls, will put a smile on your face every time!

We started out with the ends of a '32 headlight bar. Next, we cut off the feet and made a bracket out of some flat stock and a piece of square tube cut at an angle. Notice the cap on the tubing is threaded along with a flat screw in the flat stock.

With the headlight bar welded to the bracket, it mounts underneath the frame and can be removed quickly.

We like to mount the lights as low as possible. A little torch heat and some bending gives the bars a nice, graceful look.

Check out the gusset for added strength on the bottom of the bar as it disappears into the frame cover. It has one hole drilled into it and does double duty as a wire loom. Now if you were driving "this nasty-looking and sounding little bitch" around the pits during Speed Week, wouldn't you take off the headlights?

The Dash and Gauges

We feel the best look when chopping a car is to lay back the windshield posts to meet the top, rather than stretching the top with a filler piece to meet the posts. We also feel the best method is to cut the cowl all the way across, but this method tips the bottom of the dash in at the same time. With 7-plus inches cut out of the '34, it laid the posts back about 17 degrees, which made any gauges mounted in the dash face the driver's lap. A simple and pretty cool solution was to tip the gauges up and the center-mounted tach up and over.

The first step was to remove the three-hole stock gauge panel and the ashtray, and to fill all the holes. Next, the smaller gauges and switches were laid out, and a sleeve was fitted into the tach hole. After welding in the tach sleeve at both an upward and inward angle, the lead was applied.

The smaller gauges and switches also had sleeves welded in at an angle and were leaded in.

Oh-so-simple-looking and bitchin' at the same time.

Making the Complicated Look Simple

At the end of the day, it's impossible to nail down the term "hot rod," because it's really about what touches *your* emotions and stirs *your* soul.

For Keith and me, it is taking an old Ford and boiling it down to its purest form, much the way young hot rodders did when they returned to the dry lakes in Southern California after the war.

Certainly there is much more to the hot rods we build than meets the eye, but that's really the idea isn't it? Something as complicated as chopping a car 7 or 8 inches while laying back the windshield and keeping the proportions correct, to doing something as simple as removing the middle hinge on the doors and allowing your eye to run unimpeded from the front to the back, makes all the difference. It is about making the complicated look simple, unencumbered with the unnecessary details and overworked ideas that seem to plague so many cars in an attempt to make them a showcase for a builder's ideas and skills. We often hear the comment, "Your hot rods look as if they are seven-eighths scale."

While carving a bird from a block of granite, a great sculptor once said, "I just cut off everything that kept him from flying."

Introducing the George Poteet '34

AUTHOR'S NOTE: Here is the completed 1934 three-window coupe detailed in the preceding sequence. The coupe is only 54 inches tall on wheels and tires, and that is in street trim.

When I sat in the car, it was comfortable, there was plenty of legroom, and I could see out just fine. I have driven their cars quite a bit, and the only issue I ever had was that I had to bring my own peep mirror to attach to the driver's side door, which was the only real blind spot. Ken grimaced when I bolted one on his coupe for the drive out to Bonneville, but I think I actually saw him using it once, and he wanted it left on for the return trip. I will be curious how well one can see out the back slot window on the '34 while driving. Here are some photos of the "finished" '34 taken at El Mirage in June 2008. Beautiful . . . it belongs on the dry lakes or on the salt of Bonneville.

George Poteet's finished '34 Ford coupe photographed at El Mirage in June 2008.

The end of the day at El Mirage, June 2008.

ARMSTRONG

13

VERN AND KEITH TARDEL

AUTHOR'S NOTE: I have known Vern and Keith for some time now. Every time I go out to Vern's to photograph, I'm transferred to another world. It is an environment unto itself, and I love wandering around, spending the quiet visual time to discover something new and rediscover something old. Vern is an original, and now with Keith's Rex Rod and Chassis back in the fold of Vern's shop, the production value of their authentic hot rods has only improved, if that's possible. I've spent many hours photographing the shop and the surrounding area, as well as the Tardel group circus/gathering of the faithful at the Bonneville Salt Flats. I'm there every August, watching them push a car up to whatever they can get out of it during Speed Week, weather permitting. Hanging out with these guys at Bonneville is always a treat. Besides the serious side of racing, there is always a fun side to the event.

Quasi-Organized Mayhem, Inimitable Style

By Kevin Thomson

On a gently rolling country road in Sonoma County, vineyards and tree-covered hillsides give way to the occasional house. It's a peaceful scene, and on the edges it is still very rural. More than 30 years ago, Vern Tardel and his wife, Karen, bought a house in the vineyards to raise their sons, Vern Jr., Matthew, and Keith. Behind the house, in a grove of fruit trees, is a long, low, shed of a building and a smaller, barn-like structure. Racks of leaf springs and banjo rearends lay in the shade next to crates packed with carburetor carcasses. Frames, cabs, and bodies shelter fauna and foliage.

Long before Vern bought the house, he was a car-crazy teenager in love with the Norm Grabowski and Tommy Ivo rods. He'd save his money and spend it at Ed Binggeli's Bing's Speed Shop or at the wrecking yard. By the time he was a senior in high school, he had built a hot rod '25 Model T roadster that was the start of the insanity. The mid-'60s saw Vern take a detour for the biker lifestyle. He built a few bikes and rode around with a band of hellraisers known as "The Misfits." It was *Easy Rider* before the movie was even pitched. The lifestyle took its toll, until Vern realized he was headed for an early grave, jail, or both, so he quit it.

Vern went back to cars. For many years, he ran a body shop doing collision work. All the while, his interest in early Ford hot rods never let up. He began

A bird's-eye view of the action outside the Tardel place.

to amass parts—every kind of part, especially the hard-to-find items you need to actually build a hot rod by the seat of your pants. Vern is a believer in the axiom that hot rods are built with what you've got lying around, and to that end he has made sure he is in possession of what it takes to get the job done.

After 30 years of active collecting, the shop is now a masterpiece. There is no other shop quite like Vern's. No television producer or art director could even hope to come close to re-creating the quasi-organized mayhem or the inimitable style. To merely inventory what is there would do no justice whatsoever to the tableau. The shop is not a museum, although it is tempting to refer to it as one. It is a working shop and a collection that ebbs and flows with the rhythm of what is needed, what is not, what is coming in, and what is going out. It would take a team of archaeologists weeks to catalog and months more to decipher the overlapping histories on the walls alone. By the looks of things, Vern was at one time or another into beer, milk, dragsters, speedy service, cars, motorcycles, and friendship.

Vern is tight with his friends, and many have stuck with him since their high school days. Mentor Ed Binggeli is still around after all these years, actively working on the flathead eights he knows so well. Vern's childhood friend Terry Griffith might have moved a few miles north, but he still handles all the wiring on the Tardel rods. Author Mike Bishop is a frequent collaborator in the shop and on books with Vern.

This flathead-powered Ford phaeton sitting on Deuce rails belongs to Joe Fazio.

If you don't get it, well, give it another try.

Outdoor sculpture, flathead style.

Tardel serves flatheads.

Although others have moved out of state, they all converge again when it is time to run the little yellow record-holding roadster at Bonneville that Keith Tardel and the gang built. The scene at the Tardel pit at Bonneville is one of the never-a-dull-moment variety. Plenty of shade and all the comforts of home the guys can bring are laid out next to the trailer. A never-ending stream of visitors in hot rods, motorized easy chairs, and mini bikes come by to visit or witness the magic of the mayhem. It all speaks to the reputation Vern has earned and the affable man that he is.

Vern is an intense person with anywhere from 1 to 10 million different things on his active mind. New ideas are always coming around, and he is in near-constant motion, riffing off on an idea or something someone said in the shop just seconds ago. With a workload that would crush most mortals, Vern is friendly and quick with the jokes and gently self-deprecating humor. Somehow, there is always a bit of time for those who visit, either out at the salt or back home at the shop.

Despite making it look easy, Vern admits to the strain of having over a dozen cars under construction at any given time. On top of that, there is the parts line and the website and the new venture with SF Flatheads. To keep the ball on the ground, Vern hired on Roland Farwell to keep the website and shipping moving along. Out in the shop, the return of Keith Tardel's Rex Rod and Chassis has eased a lot of the burden. Keith's skill in nearly every aspect of building

A front view of the inside of Vern's shop.

a real, traditional hot rod allows Vern to focus on building the flathead motors and designing parts for the catalog.

The shop itself is equipped with all the tools necessary for the jobs at hand. Some are modern and others are in line with the era of the cars being built. One drill press stands more than 6 feet tall and once turned bits into parts destined for duty in the Pacific theater of World War II. Every available space is filled with parts crammed into boxes and overflowing onto shelves. Nuts and bolts are everywhere, in every sort of container, and often conveniently located next to a particular tool station. Of course, as Vern is fond of saying, "There's every kind but the one ya need!"

Music emanates softly from several corners, and carpets cover the shop floor to ease the burden on old bones. Somehow it all conspires to give a comfortable, greasy living room feel to the place. This description is given as an endearment, not an insult. Vern's shop is a welcoming place, brimming with enthusiasm and inspiration, and it feels as much like a home as it does a shop. It is a collision of automotive and mechanical creativity that would rival the studio of the most eccentric and productive artist. More than that, it is a shop in which any self-respecting gearhead would be happy to roll up his sleeves and get down to work.

Vern's office, just to the left of the "Clean Rest Rooms" signs, looks out on the general assembly area.

The view from inside Vern's office.

Vern with Brett Reed. Brett was working at Steve Moal's at the time this photo was taken. He is a talented hot rodder whom I've known for many years.

HELICOPTER
RIDES

HELICOPTER
RIDES
Ford
ROAD RACES
ARMSTRONG
TIRES

Destruction Derby
$3,500 PURSE
SUN. JULY 31 8 pm
SANTA ROSA
SONOMA COUNTY FAIRGROUNDS
CELEBRATING CHAMPION DRAGWAYS 21ST YEAR!
NOSTALGIA DRAGS
SUNDAY 3RD APRIL 1994
& Flathead CHALLENGE
NOSTALGIA
4TH ANNUAL
BONNEVILLE
HOT ROD REVOLUTION
www.HotRodRevolution.com
PRE-1949 CARS ONLY
WORLD'S FASTEST CHEVY-POWERED DRAGSTER
CHALLENGE
CHAMPION DRAGWAY
CHET HERB
EAT BEEF
Kansas Livestock Assn.
49th Annual
Bonneville
FLOPPY
STANLEY WANLASS
Ford's Out Front with "The Car of the Year"!
HOT RODS & MOTORC

It can be mind-boggling to consider the cars and vintage parts gathered under one roof at Tardel's.

The Kansas Livestock Association battles for wall space with vintage hot rod handbills.

A view of the chassis assembly area.

Vern has a dedicated engine-assembly room, complete with a print of Robert Williams' *Hot Rod Race* on the wall.

Keith Tardel

Some of the projects in the back of the shop. Vern had five or six '32 Ford roadster projects in progress when I was there taking these photographs. Check out the incredible selection of spare parts hanging from the ceiling.

ARMSTRONG
Rhino-Flex TIRES
ENTERPRISES
WINE COUNTRY GASES
CHASSIS

Keith Tardel built the yellow '27 XF/BFMR roadster in partnership with Larry McKenzie. It runs a 6-71 blown Merc flathead and has since been taken over by Vern. The '32 in the background is Vern's well-known and well-traveled ride. Keith's "Rex Rods" are currently working out of Vern's shop. Keith is a major talent in his own right.

A "maybe future" project at the Tardels' place is this '34 five-window Ford, seen here just outside the main front entrance to the shop.

A view of the Tardel Shop from the outside. All real patina.

Don't forget to wash your hands.

14

DALE WITHERS

THIS CHAPTER TIES IN with Chapter 9, on Bob Lick. It is through Bob that I met Dale and came to appreciate his craftsmanship and ability to work metal. Dale is one of those guys who is able to get metal so smooth as to allow a black finish.

No wonder Dale's shop is always busy. So busy, in fact, that Dale is not often able to work on his personal projects, including a 1934 Ford roadster that I've been watching for 10 years now that is still unfinished. No doubt, though, this one will be worth the wait when it's done. In fact, it is pretty damn nice right now.

I first photographed Dale's shop in 1997, when I stopped by to take a look at this '34 project. At the time, Dale was also working on a Corvette and a '55 Chevy, the turquoise frame of which you see here. (The other frame behind it is for the '34.) Also in the shop at the time was a '32 project that incorporated an Ardun conversion with a magneto and six two-barrel carburetors on a log manifold. The frame work was all traditional and gorgeous, as Dale's work usually is.

Since that visit, Dale picked up a 1932 full-fendered five-window from Bob Bauder. It was running a dual-four 409 Chevy mated to a Doug Nash five-speed tranny. Slowly, Dale has started to make it his own with some subtle reworks. I photographed the '32 along with the '34 roadster, which I can't seem to get out of my head, any more than Dale can get it out of his shop.

The '32 has recently been chopped and turned more into "Dale's car," while the '34 keeps inching closer to being on the road. Dale's work for customers, however, is constantly putting the '34 on the back burner.

Dale Withers, Portland, Oregon.

A 1957 Corvette frame competes for space with Dale's '34 build inside his garage.

I've been watching Dale's '34 roadster for some 10 years now, and still can't get it out of my head

Dale's '34 and '32.

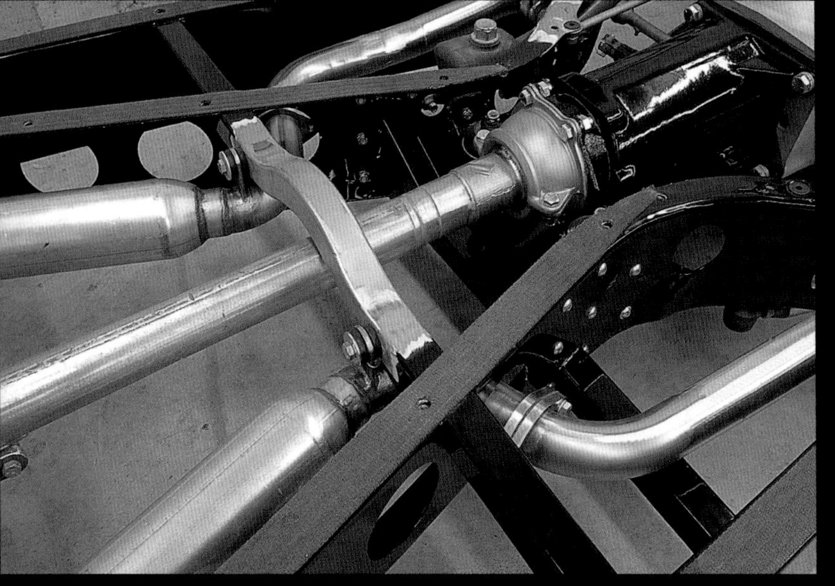

More views of Dale's '34. It features a '40 dash, and great chassis detail work. Check out the crossmember in the tranny area.

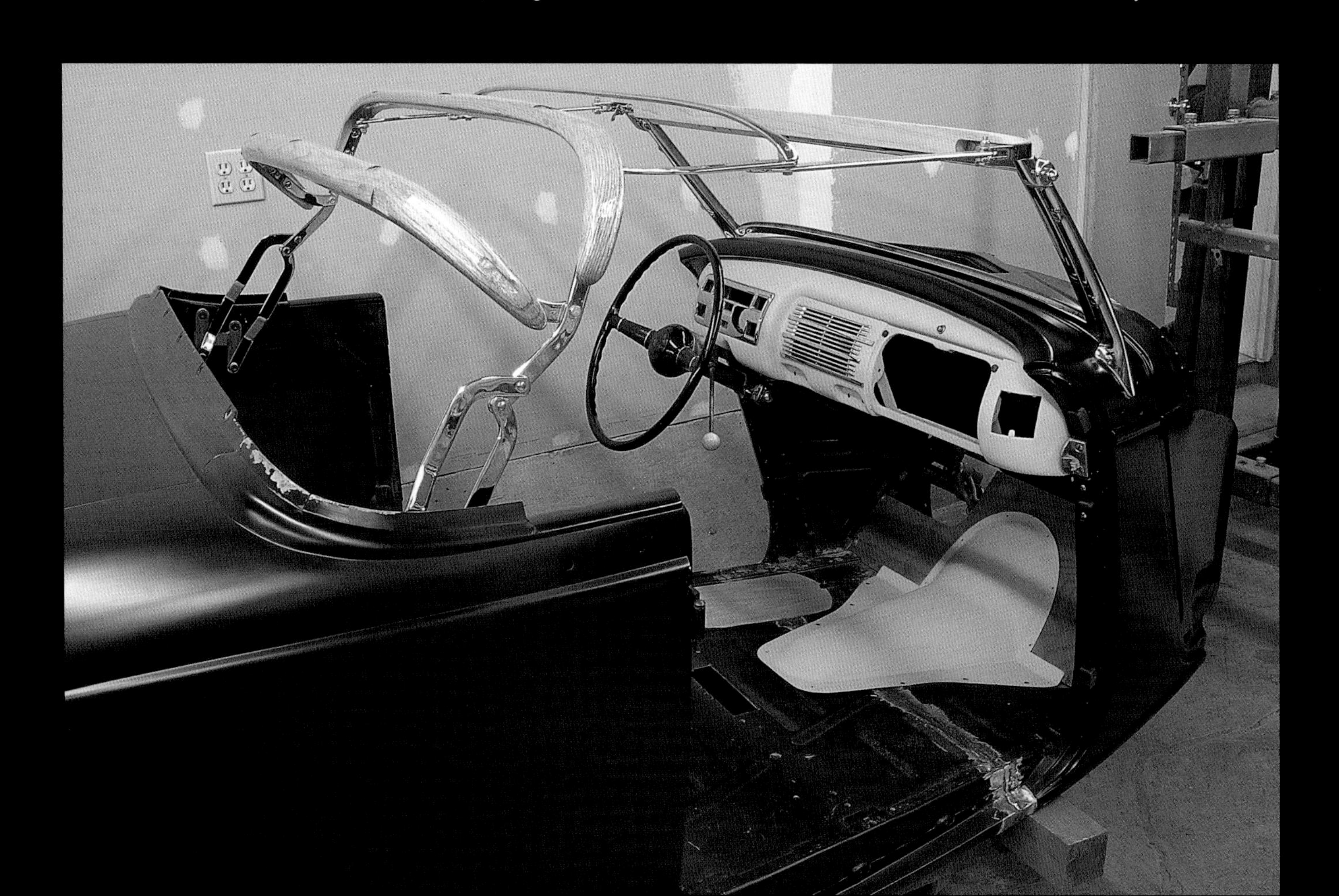

This '32 project in Dale's garage at the time of my visit involved an Ardun conversion with a magneto and six two-barrel carbs on a log manifold.

Index

MEN AND
EQUIPMENT
WORKING